Edward Coleman

A Dissertation on Suspended Respiration from Drowning, Hanging, and Suffocation

Edward Coleman

A Dissertation on Suspended Respiration from Drowning, Hanging, and Suffocation

ISBN/EAN: 9783337814335

Printed in Europe, USA, Canada, Australia, Japan

Cover: Foto ©berggeist007 / pixelio.de

A
DISSERTATION
ON
SUSPENDED RESPIRATION,
FROM
Drowning, Hanging, and Suffocation:

In which is recommended a different Mode of Treatment to any hitherto pointed out.

By EDWARD COLEMAN, Surgeon.

"Qui ſtudet optatam curſu contingere metam,
"Multa tulit fecitque puer."

Hor.

LONDON:

PRINTED FOR J. JOHNSON, ST. PAUL'S CHURCH-YARD.

MDCCXCI.

DEDICATION.

To HENRY CLINE, Esq;

LECTURER ON ANATOMY, AND SURGEON TO ST. THOMAS'S HOSPITAL.

DEAR SIR,

THAT diſtinguiſhed eminence you have ſo deſervedly attained in the medical world, and that gratitude, you might ſo juſtly claim from all your pupils, particularly from one who is indebted for his chirurgical and phyſiological knowledge, not only to your public, but private inſtructions; would, alone prove ſufficient inducements for me to addreſs theſe firſt fruits of my profeſſional ſtudies to you.

But however powerful theſe motives; allow me to add, there is another yet more cogent, and which flows more immediately from the heart.

That friendſhip with which you honoured me while reſident under your roof, and which you have kindly continued ſince I quitted that hoſpitable manſion, to enter the buſy ſcenes of life; will for ever live in my recollection, and awake the moſt grateful emotions of a feeling mind.

Permit me then to hope you will receive this Dedication as a ſmall, but ſincere teſtimony of that ſenſe I entertain of

your

your eſteem; to merit and to enjoy which, to the lateſt period of my exiſtence, is the higheſt ambition of,

Dear Sir,

your much obliged

and moſt humble ſervant,

EDWARD COLEMAN.

No. 8. FEN-COURT,
FENCHURCH STREET.

CONTENTS.

INTRODUCTION.

SECT. I.

SECT. II.

SECT. III.

SECT.

CONTENTS.

SECT. IV.

SECT. V.

SECT. VI.

SECT. VII.

SECT. VIII.

SECT. IX.

SECT.

CONTENTS.

SECT. X.

SECT. XI.

SECT. XII.

INTRO-

ERRATA.

Introduction, pa. x. line 12. after *discovered*, read *to depend on*.
Page 2. line 8. for *progress*, read *process*.
6. line 15. after *secured*, read *by ligature*.
21. in the note, after *bulk*, read *as is*.
37. line 18. for *by warmth*, read *in warmth*.
40. line 12. for *bronchial vessels*, read *bronchial arteries*.
62. line 16. after *Goodwyn* begin a fresh period.
90. line 3. after *irritability*, supply *be*.
113. line 53. for *spirits of nitre*, read *nitrous acid*.
218. line 1. for *a less degree*, read *a more gradual one*.

INTRODUCTION.

OF all the exertions of human ſkill, there is, perhaps, none which affords us more ſolid and laſting gratification, than the reſtoring to life thoſe who are apparently dead; none, ſurely, more eminently ſhews the dignity and fruitfulneſs of Philoſophy, or more clearly evinces the benefits that may be derived from the well-directed efforts of human underſtanding.

This art (if ſuch it might be called in ſo rude a condition) was, in former ages, guided chiefly by blind prejudice; the knowledge of the animal œconomy, and of life, was not ſufficiently extended, to afford maxims of any value to the practitioner; and the cauſes of death were too incorrectly marked, to ſhew, with any degree of preciſion, the means of recovery.

Accidental recoveries had, indeed, ſhewn that it was practicable; but Phyſiological ſcience was unable to explain or preſcribe the mode. It was reſerved for the eighteenth century, to exhibit, on a large ſcale, any practical ſpecimens of this mode of benevolence, and to approach, in ſome reſpect, to the ſcientific ſolution of thoſe principles by which it muſt be guided. Many ſocieties were formed on the continent of Europe, for the purpoſe of promoting this kind of knowledge; and their reports afforded the moſtmortifying reply to thoſe who had declaimed with ſuch triumph on the vanity of natural ſcience, and the impotence of human art. Their multiplied ſucceſſes, in ſo untried a path, awakened a

 general

general ardour on this ſubject, which was not a little foſtered by a cotemporary revolution in natural knowledge: I allude to the philoſophy of elaſtic fluids, which has, during the laſt part of the preſent century, received ſuch incredible acceſſions. The doctrine of *airs* was ſo intimately connected with the ſubject of reſpiration, that it could not fail to fix the attention of Philoſophers on thoſe caſes where its ſudden ſuſpenſion was the cauſe of death. It were ſuperfluous to enumerate the various theories offered by the Chymiſts and Phyſiologiſts of this recent period. Suffice it to remark, that the Humane Society of London deemed the ſubject ſo perplexed with diſcordant theories, and ſo ſuſceptible of farther experimental elucidation, that they publiſhed, in 1787, a queſtion on *the nature of the diſeaſes produced by ſubmerſion, ſuſpenſion, and noxious airs.* Two Differ-

 tations,

tations, of peculiar merit, they honoured with prizes: thoſe of Dr. Goodwyn and Mr. Kite. The ſame enlightened and benevolent body purſued this enquiry, by propoſing a queſtion—"*Whether Emetics,* "*Veneſection, or Electricity, be proper in* "*ſuſpended Animation, and under what* "*Circumſtances?*"

To this queſtion, I am about, in the following Diſſertation, to attempt an anſwer. It may be thought, that, as this queſtion is purely *practical,* any inveſtigation of the *proximate cauſe* of the malady, is ſuperfluous and impertinent, and that our views ought to be limited to the remedies employed in its cure; or, it may perhaps be ſuppoſed, that ſuch enquiry is precluded by the ſucceſsful labours of Dr. Goodwyn and Mr. Kite: but reaſon, which forbids us to abandon any thing ſo important, to blind empiriciſm; the example of theſe Gentlemen, who had from their pathology deduced their cure, and the repugnance of their inferences to each other, which countenanced a doubt reſpecting the accuracy of either,—ſeemed to prove the neceſſity of reinveſtigating, by experiment, the nature and cauſes of the diſeaſe, previous to the delineation of any plan of cure. One of theſe Gentlemen attributes death in theſe caſes to the quality of the blood in the left ſide of the heart, which has not received from the air, that ſtimulant power which ſupports the action of that organ. The other attributes it to apoplexy. I was induced, ſince the appearance of theſe eſſays, to attempt a ſeries of experiments on the ſubject, which perhaps I ſhould not have cultivated with ſo much ardour, had I not been animated by the example of Mr. Kite, from whom I received the rudiments of my medical education, and for whom, in combating his opinions, I truſt I ſhall not be deficient in that reſpect which his talents demand. Theſe experiments preſented reſults which contradicted, in many important particulars, received opinions: but I ſhould not, at ſo early a period of my life, have preſumed to offer them to the public, had I not been emboldened by the approbation of the Medical Society of London, who voted me the Humane Society's Medal. To Mr. Kite, I flatter myſelf, no apology for the freedom of my ſtrictures on his opinions, will be neceſſary. Not to have noticed his work, would have been diſreſpectful; and to have diſſembled what I found to be truth, in tenderneſs to his ſentiments, is, I am ſure, a ſacrifice that his liberality would not exact. He is

acquainted

acquainted with my experiments and conclusions, and has, I apprehend, in consequence, changed some opinions, which, in the hurry of enquiry, he had precipitately adopted.

Dr. Goodwyn has justly and ingeniously remarked, that the expression, "Suspended "Animation," is objectionable. Respiration and circulation may be suspended; but the principle of life, or the susceptibility of action, which is the source of these functions, may still remain. Life, therefore, can with no propriety be said to be suspended, when the vital principle is present. The animal must either retain the principle of life, or be absolutely and irrecoverably dead. There is no intermediate state between life and death. The distinction between the actions and powers of life, which, with so many other admirable obſervations in Phyſiology, we owe to the ingenious Mr. Hunter, clearly illuſtrates the impropriety of the language to which we object. He has proved that in many caſes, theſe powers remain when the actions are ſuſpended. The preſence of theſe powers alone conſtitute life, and form the ſole diſtinction between inanimate and animated matter. When they ceaſe to be preſent, life is *not ſuſpended*, *but deſtroyed.* Inſtead therefore of employing the term Suſpended Animation, we ſhall adopt that of *ſuſpended reſpiration*, which only ſimply expreſſes a fact, and is equally applicable to thoſe caſes which terminate in death, as to thoſe of which the event is favourable.

The neceſſity of inflicting a painful death on ſo many animals will ever be felt by minds of ſenſibility, as a cruel alloy to the pleaſure of Phyſiological reſearch. By no other mode, however, than that of experiments on living animals, can any important advance be made in this ſubject. Such experiments, in a queſtion of mere curioſity, are certainly indefenſible; but where, as in the preſent caſe, the advancement of truth conſpires with the intereſts of humanity, we muſt impoſe ſilence for a while on the remonſtrances of ſenſibility.

In the conduct of the experiments which form the baſis of the following diſſertation the moſt ſolicitous accuracy has been every where ſtudiouſly ſought.

To thoſe who are in the habits of Phyſiological experiment, nothing is more familiar than the perplexing variety and repugnance of their reſults; two experiments, though

though made in the ſame manner on the ſame order of animals, will rarely in every particular agree; for it is not only true, that different ſpecies of animals, but that different individuals of the ſame ſpecies, poſſeſs various degrees of irritability. In ſome, irritability may be excited for ſeveral hours after apparent death, others loſe it in leſs than one. The cauſe, however, of theſe variations, where they have been in any reſpect conſiderable, we have generally diſcovered ſome accidental and extrinſic circumſtance, and by multiplying and varying experiment, we have attempted to diſcriminate between what is made the foundation of *general* principles, and what is the effect of peculiar and fortuitous circumſtances. But the enthuſiaſm which we acquire in the purſuit of a favourite reſearch, and our anxiety to ſupport a cheriſhed opinion, ought ever to make the

expe-

experimental enquirer diffident of the correctneſs and impartiality of his *own* views. A bias unconſciouſly taints his judgment, againſt which the only remedy is, the vigilant eye of acute and intelligent friends, who feel more anxiety for his reputation, than tenderneſs for his prejudices; and who have no motives either to make tortured inferences, or to hide unfavourable reſults.

The ſame good fortune that has bleſſed my private life with the friendſhip of ſuch men, I have alſo eminently felt in my ſcientific purſuits. Their acuteneſs has reſcued me from my prejudices; and their aid has given me a confidence in the correctneſs of the experiments, which diſtruſt in my own individual ſkill would otherwiſe never have permitted me to entertain. I have to mention with particular gratitude, Mr. Aſtley Cooper, whoſe anatomical and phy-

physiological knowledge needs no comment; and Mr. Keir, a gentleman of distinguished ingenuity, who favoured me with his occasional assistance. And it affords me no small gratification, that my much respected friend Mr. Haighton, Teacher of Physiology, in the Borough, has made many experiments which corroborate most of the opinions here advanced.

Though submersion be the most frequent, it is by no means the only case of apparent death worthy the inquiry of the Physiologist, or the attention of the medical practitioner. Nor is the benevolent zeal of the Humane Society confined to it alone; as every case of apparent death arising from a sudden suspension of respiration, partakes equally of its bounty; and

and indeed, agreeable to this extenſive view of the ſubject, the queſtion before us is propoſed.

The ſuſpenſion of vital action from ſtrangulation and noxious airs, exhibit phenomena ſo nearly ſimilar, and require a treatment ſo ſtrictly congenial, that any inquiry into the nature of ſubmerſion, would be narrow and imperfect, unleſs illuſtrated by the inveſtigation of theſe kindred diſeaſes. To them, therefore, we have thought it expedient to extend our reſearches; and from inductions founded on a ſeries of experiments and obſervations on theſe different modes of death, we flatter ourſelves with the hope of having eſtabliſhed a general doctrine on premiſes leſs ambiguous and unſtable, than thoſe which have been the baſis of former theories.

To

To aſcertain phœnomena is the firſt duty of every inquirer into nature. We ſhall therefore, in the three firſt ſections, ſuccinctly deſcribe the uſual effects of drowning, hanging, and noxious airs. Having ſtated theſe, it will be natural to preface any enquiry into the nature of the diſeaſe, by the Phyſiology of the organs which are its ſeat; thus delineating their natural actions, before we examine their morbid condition. The Phyſiology of the heart and lungs therefore will conſtitute our fourth ſection.

Having deſcribed the phœnomena of departing life, the appearances on diſſection, and the natural ſtate of the organs concerned; we are next led to view the ſubject in a Pathological light, and to conſider that peculiar condition of the animal which forms the *proximate cauſe* of the diſeaſe. This will occupy the fifth ſection.

The

The remaining part, to which the preceding ſections are but preliminary, will be devoted to the conſideration of the cure: and in order to inveſtigate more at length, the efficacy of thoſe means which have been either ſuggeſted by ſpeculation, or ſanctioned by experience, we ſhall dedicate a ſection to each claſs; by which we ſhall be enabled to form a juſt eſtimate of their comparative efficacy and importance.

Emetics, Veneſection, Electricity alone, or combined with artificial Reſpiration, Warmth, Frictions, and Clyſters will be fairly examined by the teſts of experiment, and of reaſon; and our laſt ſection will conſiſt of concluſions drawn from the whole.

SECTION I.

On the common Effects of Drowning.

THE general effects of submersion have been described by others; and the result of our observations will be found nearly similar to that obtained by those who have already written on this subject. But, as it was necessary first to examine the appearances of animals under that circumstance, before any clear idea could be formed of the proximate cause of the disease, we shall begin with a description of the visible effects that usually arise from drowning.

As ſoon as an animal is immerſed in water, air is expelled from its lungs, and immediate attempts are made, apparently with great difficulty, to inſpire; in which a ſmall quantity of water is taken in. The animal betrays increaſing uneaſineſs; again expels air, and takes in water. The duration of this progreſs varies from one minute to four; when the muſcles of reſpiration ceaſe to act, and all ſtruggling is at an end. Some involuntary motions, however, generally ſucceed. On opening the cheſt, we find the two *venæ* cavæ, right ſinus venoſus, auricle, ventricle and pulmonary artery, loaded with blood; the left auricle nearly diſtended; the left ventricle about half; the aorta and its branches containing a quantity of blood, which, in all its appearances, reſembles venous. The lungs are diſcovered in a ſtate of collapſe, containing a ſmall quantity

of

of water, in the form of froth, but very trifling, when compared to the quantity of air expelled from the lungs, during the act of drowning. The ſtomach, on examination, preſents alſo a little water, which probably paſſed into the œſophagus when the rima glottidis was cloſed by the epiglottis; for, as the water contained in the mouth is then refuſed admittance into the trachea, it ſhould ſeem, that, at that moment, it makes its way into the ſtomach; ſo that, as ſoon as the animal attempts to inſpire, water enters the trachea; but this organ, as if conſcious of not receiving its due element, rejects the water, which is then allowed to paſs into the œſophagus. Air is again emitted, and new efforts made to inſpire, when, upon the ſame ſenſation being produced, ſimilar effects ariſe; and, after the laſt expiration, no more water enters the lungs, or ſtomach. If it were not certain, that the epiglottis clofed the rima glottidis as foon as the trachea was irritated, there would be found as much water in the lungs as the animal had expelled air; and if the ftomach or lungs continued to admit water after refpiration had ceafed, we fhould find them fully diftended, where the animal was fuffered to remain in water: but, whether our examination be made immediately after the ceafing of refpiration, or after the fpace of feveral minutes, we find no variation in the quantity.

Mr. Kite has concluded, from his experiments, that no water enters the windpipe, until the animal is dead: but the entire refult of our experiments, tends to prove the contrary; that water does get into the lungs, during the act of drowning, and that no animal, provided with lungs, can be drowned without this circumftance

taking

taking place. Indeed Dr. Goodwyn has proved this to be the fact, in a manner so convincing and satisfactory, that we need only mention, that the whole of our experiments to ascertain this point, have uniformly agreed with his.

It has been mentioned by Dr. Goodwyn, and other Physiologists, that the right auricle and ventricle are found FULL; but there seems to be some degree of impropriety in the expression, for by the term FULL is generally understood, a *cavity replete without vacuity*; and if so, the left ventricle may be said to be *full* when distended only to half its natural size, as it adapts itself to the volume of blood it contains, and in proportion as the quantity encreases, the cavity enlarges, until it attains a certain degree of distention, when it re-acts. If the heart, therefore, contain but a ſmall quantity of blood, the ſides of the auricle or ventricle will be in contact with it, and the cavity be thus obliterated. Hence we prefer the term *diſtention* to that of *fullneſs*.

The following experiments were made with a view to determine the exact proportion which the quantity of blood contained in the right ſide of the heart, bears to that in the left, after drowning.

EXPERIMENT I.

A Cat was drowned, and as ſoon as the uſual ſtruggles attending ſubmerſion had ceaſed, the cheſt was opened, the two cavæ, pulmonary artery and aorta were ſecured, and the blood contained in the heart carefully collected; upon which we found the proportions of the right to the left to be as 12 to 7.

The

The next experiment was made to aſcertain whether after the action of the heart had ceaſed, the proportions were altered.

EXPERIMENT II.

A Cat was drowned, and when the heart had ceaſed to vibrate, the two cavæ, pulmonary artery, and aorta were ſecured as before. The proportions of the right were to the left as 2 to 1.

Theſe experiments were repeated, and the quantities varied; ſometimes being as 7 to 4, at other times as 5 to 2, or as 12 to 7. So that at a medium, the proportions of the right are to the left, as about $3\frac{2}{8}$ to $1\frac{6}{8}$. The lungs were uniformly in a ſtate of collapſe.

Dr. Goodwyn has made ſome experiments to aſcertain the preciſe quantity of air contained in the lungs after death, and they were conducted in the following manner.

" A dead body of ordinary ſtature " was procured, and a cloſe compreſs ap- " plied upon the ſuperior part of the ab- " domen to fix the diaphragm in its " ſituation; a ſmall opening was then made " into the cavity of the thorax on each " ſide, and upon the moſt elevated parts. " The lungs immediately collapſed, and " water was introduced at theſe openings, " till the cavity was filled. The volume " of the water contained was 272 cubic " inches.

" The perſon on whoſe body this ex- " periment was made had been hanged.

" In

" In four ſimilar experiments, where death " was natural, he found the medium was " 169 cubic inches of air after complete " expiration."

Theſe experiments, however, are by no means concluſive, for whatever may be the cauſe of death, the animal dies with an expiration, and the tendinous part of the diaphragm is thruſt up as high as the fourth, and ſometimes as high as the fifth inferior rib; and therefore the application of a compreſs round the abdomen does not ſeem adequate to prevent the diaphragm from deſcending. Could this even be effected, as the ribs cannot be kept at any fixed point, and as the air contained in the lungs was not collected, the experiment can by no means authoriſe any legitimate concluſion.

Dr.

Dr. Goodwyn obſerves, that atmoſpheric air by means of its gravity, will enter into the cheſt, and by its preſſure on the external ſurface of the lungs, oblige them to collapſe. This obſervation, we preſume, is erroneous, for according to a well-known law in hydroſtatics, air and all fluids preſs equally in every direction. However great therefore the quantity of air may be in the lungs after the laſt expiration, the preſſure of the external air cannot be ſuppoſed to aſſiſt in repelling it. This appears obvious on a common bladder inflated, which the preſſure of the external air, by no means contributes to collapſe, but in the ſame manner as the lungs, where the preſſure is equal, its evacuation will depend on its own elaſticity and weight.

Thoſe who die a natural death muſt always have a portion of air remaining after the laſt expiration, ſince the lungs cannot be thoroughly evacuated by one, but in drowning, &c. repeated expirations are made with attempts to inſpire; but the latter are ineffectual What therefore Dr. Goodwyn has advanced on this head, appears neither eſtabliſhed by argument, nor countenanced by fact. But to determine the point, the following experiment was attempted.

EXPERIMENT.

A Cat was drowned, and after the uſual ſtruggles had ceaſed, the trachea was ſecured by a ligature, the cheſt opened, and the lungs taken out. A glaſs tube divided into drachms and half drachms, by meaſure, was filled with water, and inverted

inverted in a bafon containing the fame fluid. The trachea was then placed under the tube and divided, and the lungs being preffed, not half a drachm of air efcaped. The fame lungs when diftended contained 16 drachms of air.

This experiment was feveral times repeated on different animals, and fometimes fcarce a bubble of air was collected; in no inftance was the proportion of air in infpiration to that remaining on the lungs after expiration fo fmall as 10 to 1. The Heart has frequently been obferved to contract, or more properly to vibrate, for more than two hours after refpiration was fufpended, and that from no other ftimulus but its own blood; while in other experiments the vibrations did not continue one tenth part of that time. The right fide of the heart preferves its action much longer than the

left,

left, and the auricles longer than their correſponding ventricles.

The periſtaltic motion of the inteſtines does not continue as long as the contractions of the heart, and on opening the head, the veins, as in ordinary death, are found rather diſtended, but without the leaſt appearance of extravaſation. Our next enquiries will be directed to the effects of hanging.

SECTION II.

Common Effects of Hanging.

WHEREAS, in the lungs of animals that are ſuſpended by the neck, there is always preſent a certain quantity of air; the idea has been ſuggeſted, that they poſſeſſed no power to expel it; and that, as the lungs would then be more or leſs diſtended, the diſeaſe ariſing from it, muſt differ from that produced by drowning. To aſcertain this point, the following experiment was made:

EXPERIMENT.

A dog was ſuſpended by the neck. As ſoon as the ſtruggles became violent, the fœces and urine were diſcharged. In leſs than four minutes, he ceaſed to move. The air-tube was tied, the cheſt opened, and we diſcovered the ſame appearances after hanging as after drowning; the lungs collapſed; the right ſide of the heart overloaded with blood; the left auricle not quite diſtended; the left ventricle about half. The aorta and its branches contained blood, in quantity and colour ſimilar to that from drowning.

Hence it appears, that, when an animal is ſuſpended, the muſcles of reſpiration are capable of performing their functions; nor are the muſcles of inſpiration deprived of their action: but, as the preſſure of the cord overpowers that of the external air, and cloſes the opening of the trachea, the lungs are not found expanded, but collapſed.

Our

Our next object was, to attempt afcertaining the exact quantity of air that remained after hanging.

EXPERIMENT.

A dog was hanged; and, when all ftruggle and motion had ended, a ligature was made on the trachea, in the fame manner as in the animals that were drowned: the lungs were then removed; and the orifice of the trachea being placed under the glafs tube filled with water, the ligature was taken off. On preffing the lungs, fomewhat more than a drachm of air efcaped. Thefe lungs, when inflated, contained forty-three drachms and one half of air. This experiment was often repeated; and fometimes fcarcely any air could be expreffed from the lungs. At other times, the proportions in inſpiration were, to thoſe in expiration, as 11 and 12 to 1: but, in all inſtances, the quantity of air that remained was very inconſiderable.

In our next experiment, we endeavoured to aſcertain the exact proportion of the blood in the right ſide of the heart, to that in the left, after hanging.

EXPERIMENT.

A dog was ſuſpended by the neck, till he ceaſed to move. The two cavæ, pulmonary artery, and aorta, were ſecured by ligatures; and, after a careful inſpection of the heart, it was found, that the proportion of blood in the right, was, to that in the left, as 2 to 1.

The ſame experiment was repeated on a cat, and the proportion as 5 to 3. On a repetition of theſe experiments, it appeared that in ſome the proportions were as 9 to 4; in others as 5 to 3, and as 7 to 4. So that the medium ſtands as 2 $\frac{7}{8}$ to 1$\frac{4}{8}$.

The contractions of the heart and the periſtaltic motion of the inteſtines continue nearly as long after hanging, as after drowning, the veins of the pia mater ſeem more diſtended, but without any extravaſation.

SECTION

SECTION III.

Common Effects of Noxious Airs.

IT has been generally ſuppoſed, that when animals were immerſed in any air unfit for reſpiration, it was both taken into their lungs, and again expelled. During which proceſs a deleterious effect was produced on the ſyſtem that terminated in death.

This ſuppoſition is, however, ſupported neither by argument, experiment, nor analogy; for we find the lungs equally *collapſed* in thoſe animals deſtroyed by noxious air, as in thoſe which have been drowned. In both caſes, the firſt expiration is by no means ſufficient to exhauſt the lungs.

 The

The animals attempt to inſpire ; when they become conſcious of receiving an improper element, and the epiglottis cloſes. Air continues to be expelled, and new attempts are made to inſpire, when the trachea being again irritated by the noxious air, little or none enters the lungs, and after the laſt expiration they admit no more.

In order to diſcover the preciſe quantity of air now retained, we made the following experiment.

EXPERIMENT.

A Kitten was immerſed in nitrous air, and when it had ceaſed to breathe, the trachea was ſecured, and the lungs removed. The air was then collected as before, in the glaſs tube ; and it amounted only to $\frac{1}{2}$ a drachm ;

drachm*; whereas, in the diſtended ſtate, theſe lungs contained 14 drachms and $\frac{1}{4}$.

In the repetition of this experiment, different kinds of impure air were employed; and the proportion of it in the diſtended, to that of the collapſed ſtate was generally as 40 or 50 to 1; but in every inſtance the quantity of remaining air was very inconſiderable.

Our next object was to determine the exact quantity of blood in the right and left ſides of the heart.

To aſcertain this, we repeated the following experiment.

* We here mean the ſame bulk occupied by half a drachm of water.

EXPERIMENT.

A Rabbit was deſtroyed by nitrous air; after which, the two cavæ, pulmonary artery and aorta, were ſecured. The blood, in the right and left heart, was then collected. The proportion of the former, was to that of the latter, not quite as 3 to 2.

From a repetition of this experiment we learned however, that the proportion was ſometimes not ſo much as 3 to 2. In one inſtance, it was more than 2 to 1.

As a medium, therefore, the quantity of blood contained in the right, may be to that in the left as 5 to 3, or as 1 $\frac{2}{8}$ to $\frac{6}{8}$.

We here alſo remarked, that the irritability of the heart continues but little longer

longer than the periſtaltic motion of the inteſtines, and that in theſe experiments, they both ceaſed ſooner than in animals deſtroyed by drowning or hanging. Nor was this irritability in any one inſtance manifeſt from artificial ſtimuli after reſpiration had been ſuſpended one hour and five minutes. In ſome rabbits deſtroyed by nitrous air, the heart ceaſed to contract, from its own ſtimulus, in leſs than four minutes.

From the uniformity of theſe effects, we are authorized to conclude, that the air, in which the animals were immerſed contributed to deſtroy their irritability.

I ſhall not deny, that this effect is to be attributed to the noxious quality of the air; but ſhould rather ſuſpect, the bulk of this air, taken into the lungs of ſuffocated,

 does

does not more than equal that of the water admitted by drowned animals: for as the latter, at each inſpiration receive only a ſmall quantity of water, it is probable, the former admit only the ſame quantity of noxious air, which, mixing with what remains in the lungs, is at length nearly all expelled by repeated expirations; and a ſimilar collapſe takes place, to that which we have already obſerved, accompanies hanging and drowning.

It is a remark of Mr. Kite's, that animals deſtroyed by impure air, do not grow rigid, but remain pliant and flexible. We have however, in the courſe of our experiments, met with ſtriking examples of the contrary. Animals killed by nitrous air become ſooner rigid than thoſe deſtroyed by drowning; and in two inſtances, the rigidity of the extremities was remarkable,

even

even before the heart had ceaſed to vibrate.

On examining the head, we diſcovered ſome ſmall diſtention of the veins.

From this brief inquiry into the viſible effects ariſing from hanging, drowning, and ſuffocation, we diſcover theſe trifling variations. That in one inſtance water enters the lungs, in the other noxious air; that this air poſſeſſes a greater tendency to deſtroy the action of the heart, than either hanging or drowning, and that after the former, more blood is found in the head, though its proportions in the different ſides of the heart, are nearly equal.

The lungs in all theſe are in a ſtate of *collapſe*. Theſe conſiderations, eſpecially the *laſt*, incline me to believe, that the cauſe which

which produces death in one inſtance, operates alſo in the others. But prior to an inveſtigation of the *proximate cauſe* of the diſeaſe, a curſory examination of the phyſiology of the heart and lungs may not be improper.

SECTION

SECTION IV.

Physiology of the Lungs and Heart.

IT is by no means our design to extend this investigation to every advantage that results from respiration, as our voice, smell, &c. but merely to take a rapid survey of those functions more immediately connected with *life*.

On this subject Dr. Goodwyn has bestowed no small share of attention; and though the result of our own observations does not permit us to yield assent to many of his opinions, yet the resources of his ingenuity, and perspicuity of arrangement ever claim our admiration and applause.

But

But before we inquire into that particular connection which subsists between breathing and life, our first object is briefly to consider the manner in which respiration is performed in health.

The expansion of the thorax in ordinary inspiration is effected by the intercostal and other muscles, and its cavity lengthened by the descent of the diaphragm, but in laborious inspiration, the *serratus major anticus*, *pectorales,* &c. bear a considerable part.

Expiration is said to be both an active and passive process: it is active when the abdominal muscles compress the viscera, and draw the ribs downward and inward; and passive from some of the muscles of inspiration at this time relaxing.

The lungs themſelves are ſomewhat elaſtic; but are paſſive in reſpiration. They may not unfitly be compared to a pair of bellows, and the muſcles of reſpiration to the power that works them: in their ſtate of expanſion, or when the muſcles of inſpiration act, a cavity is formed which admits an influx of air, but when compreſſed, or by the muſcles of expiration acting, the cavity is leſſened and the air expelled. Thus, by this alternate dilatation and contraction of the thorax, the proceſs of reſpiration is ſupported.

The action of theſe muſcles in a ſtate of health is involuntary, and is leſs influenced by the will than moſt of the other muſcles in the body: we are able, however, for a ſhort interval, to check or increaſe their action, but that it ſhould not be wholly ſubſervient to our will, is very wiſely ordained; for otherwiſe the powers of reſpiration muſt ceaſe whenever the ſenſes are ſuſpended by ſleep or inſanity.

It has been generally ſuppoſed that one of the natural functions annexed to the lungs was that of aſſiſting, by their alternate dilatation and contraction, in propelling the blood from the right to the left heart, but in health they ſeem to poſſeſs no ſuch power; for if circulation depended on their mechanical action, ſuſpending our breath for one twentieth part of a minute would produce a ceſſation of the heart's motion, and we ſhould then have but one pulſation to one expiration, whereas in health we have four.

Let theory deviſe what principle it may to explain, that in health the lungs poſſeſs

no

no power of propelling the blood from one ſide of the heart to the other, the matter of fact is clear; (and it will appear ſo from an experiment contained in the next ſection) that the right ſide of the heart, unaſſiſted by the action of the lungs, is capable of ſending blood to the left, even after reſpiration has ceaſed. If then the heart, in a ſtate of debility, can perform this function independent of the lungs, can it be ſuppoſed unequal to it in the vigour of health? Groundleſs therefore is the ſuppoſition that attributes this office to the lungs in ordinary reſpiration.

But a ſubject more delicate and abſtruſe, a ſubject that of late years has been warmly and ably controverted, now claims our attention; I mean the alteration induced on the blood in the lungs, the cauſe on which this alteration depends; and what effect it produces on the animal œconomy.

To investigate the peculiar change which the air may undergo in the lungs, is but of little consequence to our present inquiry; but it is absolutely necessary to trace and ascertain the effects produced by the air on the blood, before we can obtain any knowledge of the *proximate cause* of the disease.

We are inclined to the opinion of the ingenious Dr. Crawford, that a principal advantage derived from respiration, is *animal heat*; that when the blood returns from all parts of the body to the lungs, it has lost a quantity of its latent heat *, and

* According to Locke's definition, heat is a sensible quality; and if this definition be admitted, then, properly

and imbibed ſome noxious quality; that in the lungs it meets with atmoſpheric air, containing a portion of dephlogiſticated air, which is known to poſſeſs heat in a latent form; that it abſorbs part of this heat, and at the ſame time imparts to the air which remains, its impurity.

That the blood having thus robbed the air contained in the lungs of a portion of its latent heat, and rendered that which remained ſenſibly warm, the air is expelled,

perly ſpeaking, there can be no ſuch thing as latent heat, as that muſt ceaſe to be heat when once it becomes inſenſible; but as the term appears to convey the idea we wiſh, that of a principle or quality exiſting in a body which cannot be meaſured, but under certain circumſtances can produce ſenſible heat, we have preferred it to others; and perhaps there is a greater impropriety in Locke's definition of heat, than in the term employed.

and fresh air taken in to undergo a similar process.

Dr. Crawford, in the course of his experiments, had occasion to observe that animals immersed in a warm, did not so soon phlogisticate a given quantity of air as those immersed in a cold medium; nor is the reason inevident; for when the blood arrived at the lungs, it had not lost so much of its heat, consequently did not require to rob the air of so much of its purity; whereas in the other case, the animals being immersed in the cold medium, were obliged to generate more heat; but to effect this, they must consume a greater quantity of dephlogisticated air than those in the warm medium. It is also observed, by the same author, that the difference between the colour of the venous and arterial blood, was diminished by exposing animals

animals to heat, and increaſed by expoſing them to cold.

The object of theſe experiments was to prove, that in proportion as the atmoſphere is cold, more or leſs heat is abſorbed from the air, to keep up an equilibrium of heat; and it is remarkable that the animal in the warm medium died firſt, notwithſtanding the blood was florid, and the ſurrounding air more pure than that which the animal breathed when in the cold medium.

The one dying ſooner than the other probably depended on debility; that the one in the warm medium, from being obliged to generate cold, or more properly *reſiſt heat*, was rendered weaker than the other, from this being a more expenſive proceſs to the ſyſtem than generating heat;

for there appears ſuch a tonic power in cold, that an animal will allow of its natural heat being diminiſhed ſeveral degrees without inconvenience, but cannot ſuffer its ſenſible heat to be increaſed more than ſix degrees at moſt of Fahrenheit, without death taking place. Hence it would ſeem that although the fluids of the one contained more of the ſtimulating quality than the other, yet from the ſolids not being ſo ſuſceptible of action, life could not continue ſo long: and it appears evident, that if the animal in the cold medium, could have exchanged its blood with that in the warm one, the difference in the duration of life would have been ſtill greater.

The objections adduced againſt Dr. Crawford's truly ingenious theory ſeem to poſſeſs but little weight. It is urged by ſome, that if breathing be the ſource

of

of animal heat, how can it happen that the inhabitants of the northern climates breathe no quicker than thoſe of the ſouthern; and yet nearly the ſame degree of animal heat is preſent in both? The reaſon appears obvious; there is always exiſting in the atmoſphere four or five times the quantity of pure air more than we conſume by one inſpiration; ſo that thoſe in the colder climates, although they breathe no quicker, nor take in a larger volume of air, yet they rob that air of more of its latent heat.

The cold atmoſphere, bulk for bulk, muſt be ſpecifically heavier than the warm, and, weight for weight, the bulk will be leſs; ſo that any given quantity of air, in proportion as it is diminiſhed by warmth, muſt decreaſe in volume, and vice verſa. Hence in a cold atmoſphere, although the volume of air taken in at each inſpiration be the ſame, yet in that volume a greater number of particles of air are received into the lungs; and it alſo ſeems probable, that, weight for weight, this atmoſphere ſhould contain more dephlogiſticated air than the warmer, ſince it is generally allowed that in proportion as its warmth is increaſed, it becomes a better menſtruum for foreign matter of all kinds.

Dr. Crawford ſuppoſes that heat is given out in the capillaries only; but there is reaſon to believe that heat is alſo evolved during the whole of the circulation; for in amputating a limb where the tourniquet has been for ſome time applied, the firſt blood iſſuing from an artery aſſumes a venous colour; and Mr. Hunter found, from tying up the carotid artery of an animal, that the blood became black; from which it may be concluded that the blood is capable of undergoing the ſame proceſs in the larger arteries as in the capillaries. In ordinary circulation however the change muſt be leſs in degree, from the circulation being here quicker, and a greater quantity of blood being in contact with fewer ſolids.

It ſeems alſo more than probable that the blood ſtill retains a quantity of heat in a latent form after it has paſſed through the capillaries and entered the veins, for on tying up the arm in common bleeding, the longer a ligature is applied, the darker the blood becomes; and at the concluſion of the operation its colour aſſumes nearly a florid hue, which corroborates the opinion that it may poſſeſs a conſiderable portion of latent heat, after it has entered the veins; and that this blood is capable of continuing the ſame proceſs, ſo long as it contains any heat to evolve. In fever, the venous blood is ſometimes nearly florid, and Dr. Crawford found that when animals were immerſed in a warm medium, the blood paſſed through the capillaries without undergoing the uſual change; both which circumſtances tend to prove, that the blood contains more or leſs latent heat after it has entered the veins; indeed, were it otherwiſe, the lungs themſelves could not be ſupplied with heat equal to other parts of the body; as the bronchial veſſels run chiefly to the bronchiæ, and theſe veſſels are found to be inſufficient for the nouriſhment of the lungs *. But the circulation

* It has been obſerved by Mr. Cline, whoſe accuracy as an anatomiſt it were ſuperfluous to aſſert, that thoſe inflammatory adheſions which obtain between the pleura coſtalis and the lungs, and which have acquired vaſcularity, are injectable by the pulmonary artery. On the ground of this fact, he conceives it probable, that the blood while circulating in

circulation in fishes puts this matter out of all doubt, for the heart of these animals is a single one, consisting of one auricle and one ventricle, both of which contract from the stimulus of black blood; and as the blood in the coronary vessels is of the same quality, its heat and nourishment must be kept up by that blood only which has passed through the capillaries.

Hence it is obvious, that if this black blood did not possess a quantity of latent heat, the warmth of the heart could not be supported, and the animal consequently must die: notwithstanding therefore that the blood, when it passes through the capillaries, evolves the greatest part of its heat, yet there still remains a portion of it

in the ultimate branches of the pulmonary artery, loses the venous character, and assumes the arterial one; and that in this state it is fit for the nourishment of the lungs.

 in

in a latent ſtate even after it has entered the right ſide of the heart: and however inconſiderable this may be, yet if it is equal to the demand, the temperature of the whole animal muſt be the ſame. With a view to aſcertain the comparative temperature of arterial and venous blood the following experiment was made.

EXPERIMENT.

A Dog was hanged, the ſternum immediately removed, and the lungs inflated until the blood in the left auricle became *florid.*

The contractions of the whole heart, at this time were powerful, and Mr. Hunter's thermometer being raiſed to 98° was introduced through an opening in the pericardium, and placed on the right ſide; the

mercury

mercury rose to 99° and then became stationary; it was removed to the left, and the temperature was the same; but on making an aperture into the left auricle, and thrusting the bulb down to the ventricle, the mercury fell to 97°; and on placing it in the same manner within the right ventricle, it rose above 98°.

From frequent repetitions of this experiment it uniformly resulted that although the temperature of both sides of the heart externally was equal; yet the heat of the blood in the right side exceeded that of the left, from one to two degrees.

This observation may appear rather strange, and at first seems to contradict the opinion that respiration is the source of animal heat; but the fact can be readily explained; for the blood in its passage through the

the lungs, being contained in veſſels that are in contact with air ſo much below its own temperature, the colder body muſt rob the warmer of ſo much ſenſible heat as is neceſſary to make them both equal; and the temperature of the left auricle and ventricle is kept up above that of its contents, and equal to that of the right ſide from the heat evolved by the blood in the coronary veſſels; but if the ſenſible heat of the blood in its paſſage through the lungs be diminiſhed, its latent heat is conſiderably increaſed.

If however it be a fact, that heat is imbibed from the air during the act of reſpiration, then ſhould the blood in the left ſide of the heart retain its heat longer than that in the right, when the change has taken place, though at firſt its temperature be ſomewhat inferior.

To eſtabliſh the fact the following experiment was made.

Experiment.

A Cat was ſtrangled, the cheſt immediately opened, and the lungs inflated, when the blood in the left ſide of the heart became florid; an aperture was made in the pericardium, and the mercury of a thermometer being raiſed to 99°, the temperature both of the right and left ſides of the heart was exactly 98: on opening the left and introducing the thermometer, as in the laſt experiment it fell below 97°; but on examining the right internally, it roſe to near 99°.

So far does this experiment agree with our laſt; but the temperature of the blood was re-examined fifteen minutes after, and inſtead of the right poſſeſſing two degrees of of heat more than the left, it was found, on the contrary, that the right had four degrees leſs than the left.

This experiment has been repeated by Mr. Aſtley Cooper, and in different ways, but the reſult has been invariably the ſame; that although the venous blood was ſuperior in temperature at firſt, yet before coagulation was complete, the arterial became from three to ſix degrees warmer; this, or nothing, affords a clear and deciſive proof, that heat is received by the blood from breathing; for if that blood which has paſſed through the lungs, is at firſt inferior in temperature, and ſoon after becomes ſuperior; from what can this variation ariſe but the heat received from the air in a latent form, and evolved in a ſenſible one?

We

We ſcarcely know of any animal, on whoſe blood the air does not induce ſome change, either directly or indirectly; and the great object of this change we deem to be the ſupport of animal heat; and from the maintenance of animal heat, that of *animal irritability* *.

There are animals which live in a temperature equal to that of their own; and it has been the opinion of ſome phyſiologiſts that in theſe inſtances their heat is ſupported by the ſurrounding medium. If this be ever the caſe, it probably is in aſcarides, and other animals of the ſame ſpecies, where the temperature of their medium ſcarcely ever varies; but I ſhould much

* The term irritability is very often employed in a looſe and indefinite ſenſe. We introduce it here to expreſs nothing more than a *ſuſceptibility of action*.

doubt

doubt if this is the œconomy of any animal which is placed in an element ſubject to alterations of temperature. We find that nature has very wiſely ordained, that animals ſhould poſſeſs a power of retaining their temperature for a time, whether they be exposed to exceſs of heat or cold; which in my mind is a ſatisfactory proof that their heat cannot be communicated by external temperature; indeed, if animals had not a ſource of heat within themſelves, and yet placed in an element liable to variation, life could not be ſuſtained.

It requires no great ſtrength of argument to prove that animal warmth is not produced by the ſtomach. The ſimple obſervations that, in fevers, when our ſenſible heat is greater, we take in little or no food, and that ſometimes for whole weeks; and that

the

the infant, as ſoon as reſpiration commences, and before the ſtomach receives any nouriſhment, is not leſs warm than the adult, are ſufficiently convincing that the ſtomach is not to be regarded as the ſource of animal heat.

That mere diſtention is the ordinary ſtimulus that excites the action of the heart, is the opinion embraced by ſome phyſiologiſts. Nor is it indeed improbable that a certain degree of diſtention produced by blood of a due temperature, conſtitutes the principal power which ſtimulates the heart to contract; for this power of reaction, when ſtretched beyond a certain tone, ſeems a property inherent in all muſcular fibres.

Nor do we deny that the heart, when void of blood, and ſeparated from the body, retains this aĉtion ; but this is not peculiar to the heart alone ; muſcles, whoſe natural aĉtions depend on the ſtimulus of the will, poſſeſs it likewiſe, though in an inferior degree.

That the different ſides of the heart require different ſtimuli, and that there is ſomething peculiar in florid blood, which alone is capable of exciting the left ſide to aĉtion, we cannot with Dr. Goodwyn admit.

An objeĉtion preſents itſelf, that ſtrongly militates againſt this opinion ; which is this, Why ſhould the ſame fibres, nouriſhed by the ſame veſſels, ſupplied with nerves from the ſame ſource, and *performing the ſame funĉtion*, be excited to aĉtion by different cauſes? This objeĉtion the Doĉtor is aware of, and attempts to remove it by obſerving that the animal machine offers

fers inſtances where muſcles of ſimilar ſtructure are put into action by different ſtimuli; but this is not ſaying, that muſcles *performing the ſame functions*, act from diſſimilar cauſes; which it is neceſſary to prove before any analogy can be eſtabliſhed to favour this hypotheſis.

It is far therefore from being certain that the different ſides of the heart derive their action from different ſtimuli: and let us but examine the fœtal circulation, and it will appear that both ſides of the heart contract from the ſtimulus of blood nearly of the ſame quality; that this blood is not florid in either; for even in the umbilical vein it has undergone but a very imperfect change *, if compared with that induced on the

* We take it for granted that the old opinion, of there

 exiſting

the blood which paſſes through the lungs of the adult; moreover, that the greater part of the fœtal blood arrives at the heart without paſſing through the placenta at each circulation; that is, the blood in the heart, or any other part of the adult, will receive a compleat change in the lungs, before it again returns to the ſame place; whereas, the whole of the blood in the fœtal heart will not go to the placenta, to receive the alteration at each revolution; but by far the major part will be ſent to the trunk, the head and extremities, and be returned to the two cavæ, without having entered the umbilical arteries.

exiſting an actual communication between the veſſels of the mother and child, is now exploded, as numerous experiments have been made to prove the contrary; and few at preſent adhere to that doctrine.

The

The blood in the umbilical arteries, is ſimilar to that in the trunk of the pulmonary artery of the adult circulation ; that is, it is impregnated with phlogiſton, and poſſeſſes little latent heat ; whereas that of the mother in the cells of the placenta is loaded with heat, and has little phlogiſton. In the minute branches of the arteries the change is performed ; that is, *only* ſo much latent heat is imparted to the fœtal, and ſo much phlogiſton received by the maternal blood, as is neceſſary to reſtore the equilibrium of heat and phlogiſton ; the heat therefore which is received by the fœtal blood will be ſmall, in compariſon with that imbibed by the blood of the adult in the act of reſpiration ; as *only* that quantity of heat can be imparted from the maternal to the fœtal blood, as can make both their qualities with reſpect to heat and phlogiſton *equal.*

When the fœtal blood has undergone this change, it is returned by the umbilical vein ; and part of it will paſs through the ductus venoſus into the inferior cavæ, and mix with the blood brought from the lower extremities ; but a greater part will paſs through the vena portarum to go to the liver, where, by paſſing through capillaries, it muſt aſſume the venous quality before it arrives at the right auricle ; it then unites with the blood ſent from the lower extremities and trunk of the body in the inferior cavæ, and on entering the right auricle, it mixes with the ſtream of blood coming from the head and ſuperior extremities, none of which has been to the placenta to receive the change.

The right auricle propels part of this blood (which muſt be dark) into the left, and

all

all the blood that paſſes through the capillaries in the lungs alſo enters the left, ſo that the blood which produces the contraction of the left ſide of the fœtal heart muſt be more phlogiſticated than that of the right, as part of the blood in the left auricle has paſſed through the lungs.

If the quantity of blood conveyed to the placenta by the umbilical arteries, be compared with that ſent to the head, ſuperior and inferior extremities, and trunk, it will be found that not one fifth part of the blood goes to the placenta at each revolution, nor can this blood receive but *half the heat* the maternal blood contains; moreover, as the greater part of it muſt firſt paſs through capillaries before it arrives at the heart; and as that which does not paſs through capillaries mixes with venous blood, it is obvious that both ſides of the fœtal heart contract from the ſtimulus of black blood, and that the blood of the left ſide muſt be blacker than that of the right.

From the blood in the fœtus receiving a degree of change ſo inconſiderable, when compared to that produced in the adult by the ſame proceſs, a doubt might at firſt ariſe whether in both it was deſtined to accompliſh the ſame end, the ſupport of animal heat, and from thence that of animal irritability. That it is, will appear even from ſuperficial enquiry; and that the fœtal circulation, far from invalidating, countenances the opinion which derives animal warmth from the act of reſpiration.

The experiments of Dr. Crawford have already enabled us to obſerve that the quantity of heat abſorbed in breathing is propor-

tional

tional to the temperature of the ſurrounding medium. The obſervation holds equally good in the fœtal circulation; for as the fœtus is ſurrounded by the liquor amnii and uterus of the mother, the quantity of heat carried off muſt be extremely ſmall; and that which is employed being alſo trifling, there is no occaſion for more being abſorbed than is neceſſary to ſupply the conſumption of fœtal heat; and if the whole of the fœtal blood went to the placenta at each revolution, the inevitable conſequence would be death; for the power of *reſiſting* heat muſt then be called forth to action; and this in the fœtus is very inconſiderable.

On the adult, nature has wiſely beſtowed two powers for generating cold; that of evaporation from the ſurface of the body, and a power independent of this;

but the fœtus can only poſſeſs the latter, as no evaporation can take place from the ſurface of the body; and as the fœtus is deprived of this power, and as the temperature of its ſurrounding medium, the liquor amnii, is ſo much above that of our atmoſphere, if an equal degree of heat were abſorbed in the fœtal, as in the adult circulation, the animal muſt periſh; ſince the act of reſiſting heat for a few minutes is very diſtreſſing, even where the additional power from perſpiration is preſent, to counteract its deſtructive accumulation. Admirable therefore is the proviſion which nature has made, for maintaining a proper degree of heat, both in the fœtus and adult; the former is placed in a warm medium of uniform temperature, which permits but little heat to be conſumed, and the circulation is ſo regulated as only to allow the abſorption of a ſmall and *limited* quantity of heat; ſo that

great

great powers for refifting heat are here unneceffary.

But in the adult, the varying and changeable temperature of the air makes it neceffary that more or lefs heat be abforbed, to correfpond with the variation to which it is expofed. We are therefore immerfed in an atmofphere fupplied with fufficient heat to anfwer our demand ; and by evaporation, &c. we are enabled to refift heat, fo as to prevent its undue and deftructive accumulation ; on the other hand, from the warmer medium which encompaffes the fœtus, we may gather the reafon why a fmaller portion of heat fhould be imbibed, and from this being *limited*, why it ftands in no need of evaporation for the generation of cold.

Were

Were the change induced on the blood during circulation intended *ſolely as a ſtimulus to ſupport the action of the left ſide of the heart*, then ſhould the alteration produced in the fœtus be equal in degree to that produced in the adult; but that this is not the fact we have already, and we hope not unſucceſsfully, endeavoured to prove; and indeed if this was the intention of nature, it is highly improbable ſhe would have ſo contrived it that the connection between the mother and child ſhould take place at the umbilicus, where a great part of the blood which has been at the placenta, muſt firſt have paſſed through capillaries before it enters the left auricle, and where its purity in the right ſide of the heart would be ſuperior to that in the left. We might ſooner ſuppoſe that the umbilical vein would have terminated in or near the left auricle, to ſupply it with blood thus duly altered, than

than that the blood contained in the *left side of the heart*, ſhould be *ſimilar in quality* to that in the umbilical arteries which goes to *receive* the alteration ; for, in this circumſtance, the vein contains blood that has undergone the change, but the arteries carry blood that is going to receive it.

If therefore the left ſide of the fœtal heart and the whole of the arterial ſyſtem poſſeſs no ſtimulus but that of black blood ; if the pulmonary artery in the adult be excited only by this blood ; if, in a word, the heart of fiſhes act on no other blood, is it not obvious (at leaſt as far as induction and analogies can prove) that in the adult alſo venous blood can excite the action of the *left ſide of the heart and arterial ſyſtem*, and conſequently that the two ſides of the heart do not require to be ſtimulated by diſſimilar cauſes ?

From

From conſidering that one ſide of the heart in the adult circulation contains black blood, and the other florid; and that in ſuſpended reſpiration the left ſide firſt ceaſes to act, when both contain black blood, Dr. Goodwyn, we preſume was induced to conclude that venous blood which ſupports the action of the right ſide, was an unfit ſtimulus to keep up the action of the left.

The obſervations however we have ventured in ſupport of the idea, that the whole of the heart owes its action to one and the ſame cauſe, oblige us to withhold our aſſent from that of Dr. Goodwyn; but before we attempt an explanation of the cauſe which protracts the action of the right ſide of the heart beyond that of the left, we deem it neceſſary to inſtitute a previous inveſtigation of the

effects

effects produced on the heart by blood that has been duly changed, and next enquire into the consequences that must ensue when no alteration has been given.

We have already observed that when the blood arrives at the right side of the heart, it is impregnated with phlogiston; and deprived of the greater part of its latent heat; in health it is to part with its phlogiston or inflammable principle in the lungs, and there also receive a fresh supply of heat; it is then propelled into the left side of the heart, and thence through the whole of the circulating system, to evolve and distribute heat, and receive phlogiston.

In consequence of this process, the left side of the heart and coronary vessels are supplied with blood, which distributes heat

heat and nouriſhment to the whole of the heart; and in ordinary circulation it is probable that the heart derives its heat principally from the blood in the coronary veſſels; but if the motion of the circulating fluid be checked, or totally ſuſpended, then would the blood in the cavities of the heart, continue to undergo the ſame proceſs; at leaſt ſo long as it poſſeſſed any heat in a latent form; for it has already been proved, that if blood be delayed in the larger arteries, it is known to aſſume the ſame change and appearances as when it has paſſed through capillaries. The blood within the coronary veſſels not only ſupplies the left ſide of the heart with heat, but alſo the right; and if the heart derived heat ſolely from the blood within its cavities, their temperature in health would be equal; for although the blood in the left ſide of the heart, might contain 60 degrees of latent

heat,

when the right poſſeſſed but ſix; yet if the *ſenſible heat evolved* be only equal to two, their temperatures muſt be the ſame.

The reſult of multiplied experiments authorizes the aſſertion, that immediately after the action of the left ſide of the heart is increaſed by florid blood, the right alſo becomes equally affected; nor is this effect an unnatural or unexpected conſequence; for as the coronary veſſels ſoon receive this blood, and as theſe veſſels are going to both ſides of the heart, the heat and irritability of both muſt be equally ſupported.

The great and natural ſtimulating power that keeps up the *action* of the heart, we have already ſuppoſed to be diſtention; but this muſt ceaſe to act as a ſtimulus whenever the blood becomes incapable of ſupporting the irritability of the heart, by imparting to it its wonted and neceſſary degree of heat. To effect this the blood muſt part with its inflammable principle in the lungs, and in return imbibe from the air a freſh ſupply of latent heat.

Dr. Cullen imagined, that the heart's continuing to act after breathing had ceaſed aroſe from habit; but were that the caſe, why ſhould the action of the right ſide of the heart outlive that of the left; and why ſhould not this influence of habit extend equally to arteries? Inflating the lungs ſoon after reſpiration has ceaſed, generally increaſes the action of the heart, even from the firſt expanſion; and it ſeems to ariſe from the mechanical ſtimulus, which the lungs apply to the heart by diſtention; as in proportion to the expanſion their ſurface will preſs upon the two ſides of the heart, and thus become an irritator.

If inflation however be deferred for a conſiderable time, the ſame effect will not follow; as this degree of irritability is ſeldom permanent; and diſtention of the lungs ſoon ceaſes to be an adequate ſtimulus; but by making repeated inſpirations to one compleat expiration, the irritability of the heart is ſoon revived, and an action produced by each inflation. This depends on the proceſs of circulation being duly carried on, and the neceſſary ſtimulus imparted from the air to the blood, which increaſed the living powers of the heart, and rendered it ſuſceptible of irritation from ſo ſlight an external cauſe, as the mechanical action of the lungs.

To this opinion, of the action of the heart

 proceeding

proceeding from mechanical ſtimuli, Dr. Goodwyn oppoſes this inference: If it were ſo, ſays the Doctor, any aerial fluid would be then equally effectual. But this is rather unfair reaſoning; for it is agreed on all ſides, that a change in the blood is neceſſary to the life, and uninterrupted action of the heart: and although the introduction of noxious air may prove as great a ſtimulus to the ſurface of the heart as any other, yet if the blood ceaſes to receive the change when the heart acts, the irritability of this organ muſt gradually diminiſh, as the blood continues to evolve its heat, without receiving the uſual ſupply; and what before was ſufficient to irritate the ſurface of the heart, no longer poſſeſſes that power. It is true, the heart will act on introducing any air into the lungs for one or two inſpirations, if the experiment be made immediately after breathing is ſuſpended; and

this

this is a circumſtance that corroborates the opinion of this action ariſing from mechanical ſtimuli. That it is not to be aſcribed to any change immediately induced on the blood *already in the left auricle* is obvious; for the right ſide of the heart muſt be excited to action before the left can receive blood that has undergone the change; as no alteration can be given to the blood contained in the auricle.

Dr. Goodwyn is of a contrary opinion; for he obſerves, "that the contractions of "the left auricle and ventricle are imme-"diately effected by the quality of the "blood paſſing into them."

We ſhall endeavour in the next ſection, to demonſtrate by experiment, that no alteration can be produced on the blood in the trunks of the pulmonary veins and left auricle, if the communication be cut off from the right ſide of the heart: and it muſt be manifeſt, that if the blood *already* in the left auricle could receive an immediate change, from the air in the lungs, the right, which is equally in contact with them, ſhould alſo receive it.

This opinion we are therefore diſpoſed to regard rather as one of the many off-ſprings of the author's fruitful ingenuity, advanced to ſupport a favourite hypotheſis, than to evince the genuine dictates of his judgment and conviction.

That the right ſide of the heart continues to act, after the left has ceaſed, is a phenomenon that has been noticed by almoſt every phyſiologiſt; but few, if any, have attempted to unfold its cauſe. Indeed Dr. Goodwyn

Goodwyn appears to be the only one who has ſeriouſly endeavoured to explain its rationale, and attempt its illuſtration; and though there is no authority to which we would more gladly refer, yet we cannot here adopt his opinion, *that the left auricle and ventricle, firſt ceaſe to act, from the ineptitude of venous blood to excite their contraction*; and *that this is the immediate cauſe that ſuſpends circulation in drowning, &c.*

But in order to explore the true cauſe of this phenomenon, let us once more recollect that the blood, when it arrives at the right ſide of the heart, has loſt the greater part of its latent heat; that in health it receives this ſupply in the lungs; but that *in ſuſpended reſpiration*, the blood paſſes through the minute ramifications of the pulmonary artery into the pulmonary veins, without receiving this neceſſary quality, and inſtead of diſcharging phlogiſton, and abſorbing heat, that it will continue to *evolve its heat*, and receive a new increaſe of phlogiſton.

An eſſential difference thus takes place between the blood of the two ſides of the heart; the right contains a fluid that ſtill poſſeſſes latent heat; but the left has little or none; and as the blood in the one is furniſhed with more heat to evolve than the other, its irritability of courſe muſt be greater; and the ſtimulus of diſtention is alſo predominant at the right ſide, which will conſequently ſupport the action of the one, when no effect is produced on the other.

That in ordinary circulation, both ſides of the heart might derive their heat principally from the blood in the coronary veſ-

ſels,

ſels, has already been remarked; but as this blood in ſuſpended breathing contains little or no latent heat, from having evolved it in the lungs, the heart muſt in that caſe imbibe its heat from the blood contained within the cavities; and that this proceſs can be carried on in them we have already ſhewn, ſo long at leaſt as their blood poſſeſſes latent heat to give out, and while the circulation is retarded or totally ſtopt. From which we conclude, that *if the right ſide of the heart in this diſeaſe poſſeſſed the blood of the left, and the left the blood of the right, the difference of irritability would be reverſed.*

If however, we have ſucceeded in eſtabliſhing as facts, that when the blood arrives at the right ſide of the heart it ſtill contains a portion of heat in a latent ſtate; that this blood in ſuſpended breathing continues to evolve heat in a ſenſible form ; that the inferior degree of irritability in the left ſide depends on the eſſential difference in the quality of its blood from that of the right ; that moreover this difference in quality proceeds from that of the left having been robbed of a quantity of its heat in its paſſage through the capillaries of the lungs; if, I ſay, theſe facts can be eſtabliſhed, then the temperature both of the right ſide of the heart, and its contents, ſhould be greater than that of the left in this diſeaſe.

The reſult of the two laſt experiments we have mentioned, allowed us to conclude, that both ſides of the heart externally are of the ſame temperature when the blood has received its due change from the air, though the temperature of this blood thus altered is inferior to that of venous ; and though the blood of the left ſide be at firſt lower in degree of warmth, yet its heat ſoon after becomes predominant.

The next experiment was made, to aſcertain the temperature of the two ſides of the heart, and their contents; where no change had been given to the blood.

EXPERIMENT.

A Rabbit was ſtrangled, and the cheſt being opened, a ſmall aperture was made in the pericardium, and a thermometer of Fahrenheit's ſcale was applied to the right ſide of the heart. The mercury roſe to 96°, where it remained ſtationary: it was then removed to the left, where it fell to 94°. On placing it within the right auricle, the mercury again roſe to 96°, and when applied in the ſame manner within the left, it fell ſomewhat below 94°.

This

This experiment was repeatedly made on animals that had been drowned and hanged, both without and within the heart, and there occurred a few inſtances where there was ſcarcely any difference in the temperature of the two ſides at *firſt*; but in all, the temperature both of the heart and its contents was predominant in the right, before the left ſide had entirely ceaſed to act. It appears therefore very evident, that the blood which paſſes through the lungs into the left ſide of the heart, without receiving from the air the neceſſary change, inſtead of being more tenacious of its heat than the right, on the contrary, loſes it much ſooner.

Thus we ſee the reſult of experiment ſanction and juſtify the predictions of theory, that when blood paſſes from the right ſide of the heart to the left, without having been in contact with dephlogiſticated air, to renovate its heat, it muſt evolve in its paſſage through the capillaries of the lungs what little it contained in a latent ſtate; and the left ſide being no longer ſupplied with its due nouriſhment and warmth, either from the blood in the coronary veſſels, or from that contained in its own cavities, muſt have its temperature reduced, its irritability decreaſed, and its action gradually ſuſpended, by the diminution of its ſtimulus of diſtention.

But far different is the condition of the right ſide; for although the blood in the coronary veſſels is incapable of ſupplying it with heat, yet the blood within its own cavities contains a quantity in a latent form, which it continues to evolve; thus is its irritability ſupported, and thus, by continued diſtention, is its action kept alive.

Dr.

Dr. Goodwyn having obſerved that in this diſeaſe all the cavities of the heart contain black blood, was induced to conclude that its other qualities were exactly ſimilar; but had it been conſidered that in theſe circumſtances the blood, in its paſſage through the lungs, ſuffers a deprivation of its remaining heat, without the acceſſion of a new ſupply, the cauſe whence originates the difference of irritability in the two ſides of the heart would have no longer remained obſcure, nor would the Doctor, to explain the phœnomenon, have been reduced to the ſuppoſition that the ſame muſcular fibres were excited to action by different cauſes, and that the blood of the ſame quality that ſtimulated the right ſide to contract, was incapable of producing the ſame effect on the left, but this difference would have been diſcovered to ariſe from the left having loſt a greater portion of its

heat,

heat, and its ſtimulus of diſtention being diminiſhed beyond that of the right.

The advantages derived from this property of the right ſide of the heart, which ſupports its action after that of the left is ſuſpended, ſeem to have eſcaped the notice and eluded the reſearch of phyſiologiſts, yet no proviſion of nature more deſervedly claims our admiration and enquiry; for in no department of the animal œconomy has ſhe managed a wiſer precaution for the preſervation of life, than by thus, after the laſt expiration of the animal, prolonging to the right ſide of the heart a ſtimulus and power of action ſuperior to that of the left.

Let us but ſuppoſe the reverſe, that the left had the irritability of the right, and the right the irritability of the left; as it is found neceſſary to the effecting a recovery, that the

the right ſhould firſt contract, and ſupply the left with blood, in order to excite it to action; and as the right, in this ſuppoſition, would ſoon be incapable of performing this function, we ſhould only be enabled to recover thoſe in whom the actions of life had been ſuſpended only a very ſhort time after reſpiration had ceaſed; whereas, from the right continuing to contract after the left is motionleſs, it is thus capable of propelling blood through the lungs into the left auricle, which being once reſtored by the arrival of duly prepared blood (even though it ſhould have ceaſed to act from the ſtimulus of its own) is enabled, by the freſh ſupply of this ſtimulating quality, to revive, and the action of the whole heart is encreaſed; but if the irritability of the left ſide were at firſt predominant, it would get rid of its own blood, and the feeble action of of the *right ſide* be incapable of ſupplying it with more.

Thus, at the very origin of the circulation, where the freſh ſtimulus is laſt applied, Nature, ever wiſe in her operations, has prudently placed a ſuperior degree of irritability, while in the left, where the irritability is inferior, the increaſe of ſtimulus is firſt received: nor will this be deemed the reſult of chance, if we but recall an obſervation we have already mentioned, that in the fœtal circulation, the ſtimulating quality of the blood is greater in the right ſide of the heart than that in the left, and that in the adult it is reverſed.

But although the blood, in theſe two ſtates of the animal, poſſeſs this difference of ſtimulus in the different ſides of the heart, yet, if an injury threaten the life either of the fœtus or the adult, the right ſide of the heart will be found to contain blood of a ſtimulating quality ſuperior to that of the left, and conſequently greater irritability; for let us ſuppoſe that, at the time of birth, the umbilical chord is prevented from carrying on the circulation to and from the placenta, the blood that runs to the left heart, from its being obliged previouſly to paſs through the capillaries of the lungs, is deprived of a portion of its ſtimulus: and thus, in the morbid ſtate, is the ſame proviſion made for the fœtus as for the adult, though their natural circulation be widely different.

There is reaſon to ſuſpect that in man there does not exiſt ſo much irritability as in animals of more ſimple conſtruction; for it ſeems that in the more perfect or complicated, as man, whoſe ſentient powers

are

are greateſt, the vital are leaſt; and we believe this will hold good in gradation with all the inferior animals, that, in proportion as the ſentient powers abound, the vital diminiſh, and *vice verſa*.

This is ſtrikingly exemplified in the polypus, which has been obſerved to regenerate into as many different polypi as divided into pieces; and theſe animals have neither brain nor ſpinal marrow.

It appears therefore not improbable to be the intention of the great Creator, that thoſe animals, whoſe powers for perceiving danger are leſs acute, ſhould be capable of receiving greater injuries without the deſtruction of life, than thoſe that are armed with this faculty in a ſuperior degree.

All impreſſions made upon ſuperior animals are immediately conveyed to the brain, and this being the great ſenſorium, the whole animal receives the alarm, and an immediate effort is made to remove the cauſe of the injury. But inferior animals, that are unprovided with nerves and brain, that are conſequently deſtitute of ſenſation, and whoſe powers of inſtinct are but feeble, Nature, we find, to compenſate for this want of ſenſation, has enabled them to withſtand injuries to a greater degree than thoſe that are furniſhed both with brain and nerves. Animals alſo that are endowed with ſuperior ſagacity, poſſeſs but a ſmall degree of irritability; and it ſeems to be juſtly remarked, that the irritability of animals decreaſes as they advance in age. This was certainly intended for the ſame excellent purpoſe, that of ſupplying the defect of ſagacity while young; but when the ſentient

ſentient powers became adequate to the neceſſity, this exquiſite irritability, which was ſo wiſely beſtowed on them while young, is no longer required.

In different ſpecies of animals, we have ſometimes obſerved that after reſpiration is ſuſpended, from drowning, &c. &c. ſcarce any action remained in the right ſide of the heart; but in ſeveral experiments, particularly in one, the cauſe of this phœnomenon we diſcovered to ariſe from an over diſtention of the right auricle and ventricle; for when a ſmall puncture was made in the ſuperior cava, and a portion of the blood contained in the right heart expelled, its contraction became extremely powerful.

Here then was indirect debility brought on from over diſtention; and there is reaſon to ſuſpect that this may frequently happen from the method of recovery ufually adopted.

There remains a fufceptibility of action in almoft every part of the body, for fome time after the fufpenfion of the fentient powers; but as animals, whatever may be the caufe of their deftruction, begin to die firft at the extreme and exterior parts; fo, in fufpended refpiration, from drowning, &c. we find the irritability of the heart outlives that of any other part of the body. One exception indeed has occurred, where the heart and extremities ceafed to act nearly at the fame time.

From confidering the length of time the heart may be made to contract after breathing has ceafed, there can fcarce be any doubt, if electricity be unable to excite it to action, but that life is irrecoverably loft; for,

for, with Mr. Hunter, we imagine life and the power of action to be intimately connected. If therefore we are incapable of calling forth this power into action, by the ſtimulus of electricity applied to the heart, there does not remain the moſt diſtant probability that the effect can be produced by the application of any other ſtimulus.

In our attempts, however, to reſtore the life of the apparently dead, we are furniſhed with no criterion for determining when this power of action is thoroughly extinct; for the exterior parts may have loſt this degree of irritability, and the heart ſtill retain it. In ſome inſtances, the heart of young animals has been made to act by electricity from ten to fourteen hours; and a gentleman, on whoſe veracity I can rely, has informed me he has ſeen it contract even twenty hours after reſpiration was ſtopped, and which is many hours longer than we have been able to excite action in any external part.

It has been obſerved by Mr. Kite, "that "the electrical ſhock is to be admitted as "the teſt or diſcriminating characteriſtic of "any remains of animal life, and ſo long as "that produces action, may the perſon be "ſaid to be in a recoverable ſtate; but when "that effect has ceaſed, there can no doubt "remain of the party being abſolutely and "poſitively dead."

With the deference due to Mr. Kite's authority, we cannot but withhold our aſſent from this opinion, ſince it appears to be fraught with ſuch imminent danger; for if we conclude that life is departed when no external action can be excited by electricity, we ſhall frequently neglect the application

of

of remedies, when the power of action an life are ſtill preſent in the heart.

There have been caſes, and I myſelf have ſeen one, where no recovery was effected, even when contractions were produced externally; but the want of ſucceſs in this inſtance is not to be attributed to the weaker powers of the heart, but to the inſufficiency of the plan of treatment; for it is probable that a recovery is not only to be effected in moſt inſtances where external contractions are viſible, but in many where this degree of irritability is deſtroyed, if *proper remedies* are had recourſe to. It appears ſomewhat extraordinary that Mr. Kite ſhould have recommended ſo dangerous a prognoſtic (built merely on hypotheſis) as that life was abſent when external irritability was not manifeſted by electricity; for it is obſerved in the ſame ſection, "that

" irritability and vital heat appear to be co-
" equal:" which opinion is incompatible with the other; for if heat and irritability co-operate, then, as external heat diminiſhes quicker than internal, it muſt follow, according to the author's own reaſoning, that external irritability muſt ſooner ceaſe than internal; and, as internal excitement may not produce external action, the concluſion that life is extinct, when irritability is no longer viſible from electricity, muſt be fallacious.

We were at firſt inclined to the opinion that irritability and animal heat might co-exiſt; that, from the latter being preſent or abſent to a certain but unknown degree, we might be able to draw a prognoſtic of the preſence or extinction of the other; but ſubſequent obſervations diſcovered this theory of Mr. Kite's to be likewiſe erroneous; for, as there are few whofe folids are not very differently excited to action by the fame caufe, fo the quantity of heat evolved from the blood, that would fupport irritability in the one, would produce no effect on the other.

This opinion is confirmed by the following experiments:

Experiment.

A fmall Puppy was drowned, and on examining the temperature of the two fides of the heart in the pericardium, the right was 98°, the left 96°. The right fide of the heart continued to act for more than two hours; and during the laft ten minutes, its temperature was 60°, that of the left 57°; the warmth of the air in the room 55°.

EXPERIMENT.

A full-grown Dog was hanged, the pericardium opened, and the temperature of the right ſide of the heart was 100°, the left 99°. The right continued to act not quite ten minutes, when its warmth was 90°, that of the left 87° and one-half: the temperature of the room was alſo 55°.

Here then action continued in the one more than twelve times longer than in the other, though with a degree of heat much inferior. We here alſo had a farther opportunity of being convinced that heat and irritability do not always co-exiſt, from the bodies of two perſons that had been executed. A powerful electrical ſhock was given, without producing the ſmalleſt external action, although three hours after execution

execution the temperature of one was 80° externally, and the other 82° at the expiration of two hours and one-half.

This ſuperior degree of heat, above that of the atmoſphere, does not proceed, as Mr. Kite imagines, from the preſence of ſome "internal animating principle;" for the longer or ſhorter continuance of ſenſible heat of any animal muſt always be proportionate to the quantity of latent heat the blood contains, and the temperature of the ſurrounding medium; whereas the difference of irritability much more depends on the readineſs with which the ſolids act when this ſtimulus is applied, than on the quantity of heat that is evolved.

Why the fibres of one animal of the ſame ſpecies ſhould more readily act than thoſe of another, from the ſame cauſe, and how we we are to difcover the different degrees of this fufceptibility of action in each particular animal, is a queftion not lefs important, than intricate to unravel. As we have endeavoured to prove that *heat and irritability* do not neceffarily co-exift, this may at firft feem to militate againft the opinion of heat being effential to the fupport of irritability; but in reality it does not, for altho' the fibres of one animal fhall act with its temperature at 60°, the fibres of another fhall ceafe with its temperature at 90°: yet this only proves that the folids of the one act from a flighter caufe than thofe of the other, and not that the ftimulus of heat is wanting. A certain quantity of inebriating liquor fhall produce violent effects on one perfon, when a much greater quantity fhall have no effect upon another.

The

The ſame reaſoning holds good in theſe experiments; for although the heat of one animal may exceed that of another, and where the inferior degree of heat is preſent, the greater effect be produced; yet the *ſtimulus* in *quality* is the *ſame*, and the difference of action depends on the moving powers of the one being more readily excited to act than thoſe of the other. Nevertheleſs, though no deciſive prognoſtic can be drawn of the preſence of *irritability*, from the preſence of *any known degree of heat*, yet the nearer the degree of heat of any particular animal approaches to its ſtandard, the greater muſt be its irritability; but it will ever be better to fix no criterion of life, and make uſe of every poſſible means of recovery, in every inſtance, than to form a hazardous prognoſtic, that may prove fatal to hundreds.

Having now examined the common effects that arise from the suspension of respiration in Drowning, Hanging, and Suffocation, and particularized the advantages derived from the Heart and Lungs, we shall, in the next Section, endeavour to ascertain the *immediate cause* of the disease.

SECTION

SECTION V.

An attempt to aſcertain the proximate cauſe of the diſeaſe produced by Submerſion, Strangulation, and Suffocation.

TO inveſtigate and eſtabliſh the proximate cauſe of the diſeaſe ariſing in ſuſpended reſpiration from drowning, hanging, &c. is a taſk that has engroſſed the attention, and exerciſed the pens of ſeveral eminent phyſiologiſts; but there has been little coincidence of opinion, each ſeeming to have ſtarted, and embraced an hypotheſis of his own.

It has been the idea of ſome, that the air contained in the lungs becomes highly phlogiſticated, and that from its deleterious

 influence,

influence, originates the diſeaſe. Others attribute it to a congeſtion of blood formed in the heart and lungs, while another claſs ſuppoſe death to be produced by apoplexy.

To none of theſe opinions does Dr. Goodwyn incline; to him it appears that from the privation of the uſual ſtimulus ſupplied by the air, the blood *contained in the left auricle and ventricle* is rendered incapable of exciting their contraction; and hence he derives the *immediate cauſe of the ſuſpended circulation.*

From an authority we ſo highly reſpect, it is with diffidence we diſſent; but argument, obſervation, and experiment all tend to prove this opinion erroneous.

If the preſence of black blood in the left heart was the *proximate cauſe* of circulation ceaſing, then we ſhould certainly find it *fully diſtended* from the action of the right, but we have endeavoured to prove that this is by no means the fact; and indeed, if the left auricle and ventricle were *fully diſtended*, and it were neceſſary for the reſtoration of life that the blood *already contained* in the left auricle ſhould undergo a change, before it was enabled to empty itſelf, then every animal would be irrecoverable as ſoon as this black blood had once diſtended the auricle; for we can appeal to the teſt of experiment to prove, that no alteration can be produced on the quality of the blood contained in the trunks of the pulmonary veins and left auricle.

To aſcertain if any ſuch change could be effected, the following experiment was made.

EXPERIMENT.

A Dog was ſuſpended by the neck until he ceaſed to move; on opening the cheſt, both ſides of the heart were obſerved to contract; but the left ceaſed in eight minutes, while the right continued to act ſtrongly. The pulmonary artery being carefully ſeparated from the aorta, and ſecured by ligature, we proceeded to inflation, which was continued fifteen minutes, without enabling us to empty the trunks of the pulmonary veins and left auricle, or produce any apparent alteration on the quality of the blood.

This experiment was repeated on a cat, during the action of the left ſide of the heart, which became leſs diſtended, but

no

no alteration in the colour of its blood could be produced. The change therefore which the blood undergoes in its paſſage through the lungs, is effected before it enters the *trunks of the pulmonary veins and left auricle*, and as the air cannot come in contact with *this blood* to produce any chymical alteration, it muſt be propelled through the ſyſtem unaltered, whenever an animal recovers; for ſuppoſing the blood within the lungs to have undergone its uſual change from inflation, as the trunks of the pulmonary veins and left auricle are here underſtood to be *full*; and as *this blood* can receive no chemical change, the left auricle muſt act on its *black blood*, and receive the contents from the trunks of the pulmonary veins (which we have ſaid has not undergone the change) before the left heart can contain blood duly prepared by the air. We were, at firſt induced to believe that the collapſe of the lungs after inflation, might have the power to empty the left auricle mechanically, by propelling the contents of the pulmonary veins onward, and by the preſſure thus applied from without, to the blood within the auricle, to ſtimulate its muſcular fibres to react, and ſo expel a portion of its contents. But there ſeems an objection to this mode of reaſoning; for if the lungs by their collapſe had any ſuch power, they muſt have exerted it at the laſt expiration, and then thoſe veſſels which are affected by this action would be ſo far emptied as to require a freſh ſupply of blood from the right ſide of the heart, before the lungs could by their collapſe, have any mechanical effect on their contents; and the next experiment proves, that after reſpiration is ſuſpended, very little blood is left within the lungs.

EXPERIMENT.

A Cat was drowned, and when all motion had ceaſed, we opened the cheſt and ſecured the pulmonary artery. A ſmall ligature was then paſſed round the trunks of the pulmonary veins, as they enter the left auricle, and both auricle and ventricle were then opened; the blood being all taken up by a ſponge, the trunks of the pulmonary veins were divided, and on preſſing the lungs very little blood eſcaped, except that contained in the trunks. The repetition of this experiment afforded the ſame reſult. We muſt therefore look elſewhere for reaſons to account for the action of the left auricle in recovery, as experiment proves that by inflation we can produce no chymical change within the trunks and auricle, nor by the mechanical action of the lungs empty the trunks, if the communication be cut off from the right side of the heart; as this, I say, cannot be effected, it would seem that when the right side of the heart acts during inflation, there is a quantity of blood sent within the lungs; and this contraction, assisted by an artificial collapse * of the lungs, propels a portion of the contents of the pulmonary veins onward, and thus produces such a vis-a-tergo on the blood within the auricle, as to excite it to contract. It has been before observed that the right side of the heart *in health* performs this function independent of any mechanical action of the lungs, and it is

* By artificial collapse we mean emptying the lungs of the greater part of their air, which will compress and evacuate the pulmonary vessels; but collapse from an ordinary expiration has no such effect.

likewise

likewiſe capable of doing it for ſome minutes after reſpiration is ſuſpended; but where the contraction of this organ is inſufficient to propel blood through the lungs, producing an *artificial collapſe* will have the ſame effect. This however can only happen where a freſh ſupply of blood has been produced by the contraction of the right ſide of the heart; for experiment demonſtrated that the quantity of blood remaining in the lungs was too ſmall to enable their mechanical action to have any effect on their contents.

It has been mentioned by Haller and other able Phyſiologiſts, that where the lungs are collapſed, an obſtruction to the paſſage of the blood through them will be the conſequence; but they have not proved *that the lungs are in ſuch a ſtate of collapſe in Drowning, Hanging and Suffocation.*

We

We have endeavoured to ſhew that Dr. Goodwyn's experiments to determine this point were objectionable, and our enquiries preſented reſults very oppoſite to his, that inſtead of the lungs being diſtended that they were collapſed, and contained but very little air. In order, however, to prove that *this degree of collapſe* was ſufficient to produce a mechanical obſtruction in the lungs in Hanging, Drowning, &c. we compared the quantities of blood in the different ſides of the heart, where the collapſe was removed to that where the collapſe exiſted.

The experiments were conducted in the following manner.

EXPERIMENT.

A Dog was ſuſpended by the neck, and in leſs than a minute the fæces and urine were diſcharged; his ſtruggles continued for little more than three minutes, when he ceaſed to move; the trachea was then laid bare, and divided, and the lungs fully diſtended with warm water (about blood heat) through the medium of a funnel; the trachea being ſecured ſo as to permit no water to eſcape, the cheſt was opened, and, contrary to all experiments made before, there was found a much leſs quantity of blood in the right ſinus venoſus, auricle, ventricle, and pulmonary artery, than in the left, which was loaded with blood, part coagulated, and the whole quite black. The experiment was repeated, and yielded nearly the ſame reſult, with this variation, that the

the right ſide of the heart had a little more blood than before, but the left was again fully diſtended.

It then appeared evident, that if by an artificial diſtention of the lungs only, without the admiſſion of air to produce any chymical change on the blood, the right ſide of the heart was capable of diſtending the left, and of expelling a part of its own contents, that in ſuſpended reſpiration there exiſts ſuch *a mechanical obſtruction in the * interior pulmonary veſſels from collapſe of the lungs*, as prevents the right ſide of the heart from getting rid of its contents.

* By interior pulmonary veſſels is meant thoſe that ramify within the lungs, and are influenced by the air; and by the trunks we mean thoſe veſſels that ariſe from the auricle, and are attached to the ſurface of the lungs.

The experiment was therefore repeated with ſome alteration.

EXPERIMENT.

A Cat was drowned, and after the ceſſation of all ſtruggles, an aperture was made in the trachea, and the lungs diſtended with air which was retained. On opening the heart we found the contents of the left ſide were to that of the right as five to four.

EXPERIMENT.

A Dog was drowned; when he ceaſed to move, cold water was introduced into the lungs. On examining the heart we found the proportions of the blood in the left were to that in the right as ſix to five.

These

Theſe experiments were repeated, and ſometimes the proportions were as ſix are to four; but in one, where the irritability was trifling, the blood was a little predominant in the right. On the contrary, in another, where great irritability was preſent, the proportions were as two to one.

It may be urged by ſome as an objection to the above experiments, that water may act as a ſtimulus to the pulmonary veſſels, ſo as to excite them to act; but it has been obſerved, that there remains very little blood within the lungs after the laſt expiration; and if water acted on them as a ſtimulus, it could not however produce any effect on the trunk of the pulmonary artery, right auricle and ventricle, which we find in part emptied.

We have obſerved that animals under the common method of ſuſpenſion, retain the power of expelling air from the lungs; but it was found not impoſſible ſo compleatly to compreſs the trachea, as to prevent any air from eſcaping: with this view the following experiment was tried.

EXPERIMENT.

The trachea of a Kitten was laid bare, and a ligature paſſed round it, that the whole of the air might be confined within the lungs. The animal ceaſed to move in four minutes and a half; and on opening the heart we found the proportions of blood in the left ſide, were to that of the right as nine to ſeven.

The ſame experiment was repeated on a Rabbit, and the proportions were as eight to ſeven.

In theſe experiments therefore, where the muſcles of expiration had not ſufficient power to overcome the compreſſure of the cord, and expel air from the lungs, the blood accumulates to a greater quantity in the left ſide of the heart, becauſe no collapſe takes place, and conſequently no obſtruction to the paſſage of the blood through the lungs.

The next experiment was made on an animal that had been ſuffocated, by diſtending its lungs with nitrous air.

In order to perform this experiment a common bladder was procured, and a pipe affixed

affixed to its neck, ſmall enough to be inſerted into the trachia of a rabbit. This pipe was introduced through a cork adapted to the ſize of a wide-mouthed bottle, which contained copper with diluted ſpirits of nitre. The nitrous air ariſing from this ſolution, was collected in the bladder, and when a ſufficient quantity was obtained, we attempted the following experiment.

EXPERIMENT.

A ſmall Rabbit was deſtroyed in nitrous air, and as ſoon as it diſcontinued to expire air from its lungs, we removed it from the medium in which it was plunged. A ſmall aperture was then made in the trachea, the bladder taken from the bottle containing the nitrous air, and the pipe introduced into the trachea in order to diſtend the lungs; which being effected, the air was prevented from efcaping, by tying the trachea. On examining the heart, the proportion of blood in the left was to that in the right as feven to fix.

The experiment was again repeated by deftroying an animal in fixed air, and diftending the lungs with nitrous air; and the proportions in the left werc to thofe in the right as thirteen to twelve.

But thefe laft experiments did not always favour our expectations, a larger portion of blood being found in the right fide of the heart, from the flight degree of irritability that remains after refpiration had been ftopt by noxious air.

Our next attempt was to afcertain if more blood were found in the lungs of an animal whoſe reſpiration was ſuſpended, and then the collapſe removed by a fluid; than where this ſuſpenſion took place without the removal of the collapſe.

We could deviſe no method to enable us to eſtabliſh this point with accuracy, but ventured however on the following experiment.

EXPERIMENT.

A Rabbit was drowned, and the lungs immediately diſtended with air; after tying up the trachea the cheſt was opened, the pulmonary artery and aorta ſecured, as alſo the trunks of the pulmonary veins. The left ſide of the heart was then opened, the blood removed, and pulmonary veins divided, the ligature was taken from the trachea, and the air expressed from the lungs. A large quantity of blood flowed from the pulmonary veins, and in a few minutes, by alternate expansion and collapse, the lungs were emptied of their contents. No accurate comparison however could be drawn between the quantity of blood present in this experiment, and that which they contained in the collapsed state; but it was evidently less in the latter, which tends to confirm the opinion of the collapse of the lungs preventing a free circulation through them; for if more blood is found when they are distended than when collapsed, this it would seem must arise from the presence of an obstruction in the one instance, and its removal in the other.

These, together with the former experiments, conspire to prove that the collapſe forms an impediment to the circulation; for if in an animal that is drowned, hanged or ſuffocated, the blood be found to predominate in the right ſide of the heart, while in another deſtroyed by the ſame means the contrary takes place merely from the introduction of a fluid into the lungs which can have no chymical effect on the blood; from what can this variation and difference of quantity originate, if not from the mechanical obſtruction in the firſt caſe, and its removal in the ſecond?

It ſhould however be obſerved that although repeated experiments prove mechanical obſtruction to exiſt in ſuſpended breathing; yet it muſt be confeſſed that the right ſide of the heart is capable of overcoming in ſome meaſure, this obſtruction, at leaſt for ſome little time after respiration has ceased, and the left of getting rid of *its black blood*; an opinion that is strongly countenanced by the following experiments.

EXPERIMENT,

A Kitten was drowned, the chest immediately opened, and the aorta secured, without including the pulmonary artery; when the heart had ceased to contract, the quantity of blood in both its sides was examined, and it was found that the left contained nearly as much as the right.

This experiment was frequently repeated, and sometimes the quantity of the blood was greater in the left side of the heart than in the right; but in all the experiments the disproportion was lessened by tying up the aorta.

In

In the animals therefore ſubjected to theſe experiments, the blood muſt have paſſed through the lungs in the collapſed ſtate; and if no ligature had been applied, this *black blood* would have been propelled into the aorta, ſince the period of examination of the heart after reſpiration has ceaſed, makes no alteration in the proportions.

Theſe experiments afford a reſult in direct contradiction to the opinion ſupported by Dr. Goodwyn, that the left ſide of the heart is incapable of acting from the ſtimulus of black blood: for they prove that whenever the right ſide of the heart is capable of ſending blood through the lungs in the collapſed ſtate, the left is alſo enabled to contract from the ſtimulus of *black blood*.

The ſame experiments may alſo at firſt ſeem to invalidate the opinion that ſuppoſes the preſence of collapſe. But every appearance of objection will vaniſh, if we but reflect that whenever the right ſide of the heart has the power of propelling blood through the lungs in the collapſed ſtate, the quantity is ſo ſmall that it can produce no effect; for we find the lungs contain but very little air, and conſequently under this diſeaſe are nearly in the ſame ſtate as the fœtal lungs; but as only a ſmall quantity of blood in the healthy ſtate of the fœtus, can be propelled through that viſcus, it appears that the blood paſſing through it during the collapſe in the adult, would not be ſufficient for the demand, as very little more blood can be ſent through the lungs after the laſt expiration, than in the fœtal circulation; with this material difference however, that in the latter a change has been given to the blood (in the pla-centa) while in the other it can receive none.

Now as the left ſide of the heart ſoon ceaſes to poſſeſs a ſtimulus that can enable it to diſcharge its contents; ſo alſo the right can no longer propel blood through the lungs in their *contracted ſtate:* for if the right ſide of the heart continued to ſend blood through the lungs when the left was incapable of getting rid of its own, we ſhould then find the blood pre-dominate in quantity in the left.

Were Dr. Goodwyn's aſſertion true, that after the laſt expiration in drowning, &c. &c. the lungs contain a greater quantity of air than in *hydrops pectoris*, then an objection would ariſe to the ſuppoſition of their collapſe forming an impediment

to

to the free paſſage of the blood; but of the experiments which he imagined authoriſed this concluſion, we have already attempted to detect the inſufficiency.

It muſt however be confeſſed, that Dr. Goodwyn's experiments ſeem ſo ingeniouſly deviſed, and the concluſions drawn from them ſo ſpecious, that at firſt they ſuſpended inquiry; and it was only by ſubſequent examination that we were able to detect the fallacy of thoſe particular ones, which he adduces to aſcertain the quantity of air remaining in the lungs after the laſt expiration. But by purſuing a mode of enquiry different to his, we obtained a reſult extremely unfavourable, and indeed contradictory to his concluſion, viz. that inſtead of the lungs containing a large quantity of air after drowning, hanging, or ſuffocation, the reſiduum is very inconſiderable,

ſiderable, and they are found in a ſtate of collapſe.

To this concluſion ſucceeded an obvious reflection, that if the circulation could be properly carried on during a *collapſe* of the lungs, why ſhould the fœtal circulation differ from that of the adult? and indeed it appears evidently to be the intention of Nature, that only a ſmall portion of blood ſhould ever paſs through the lungs in their ſtate of *collapſe*, for ſhe, ever uniform as wiſe in her operations, would never have provided a different circulation for the fœtus, if the veſſels of its lungs could have admitted through them a free and uninterrupted paſſage to the blood; but as a collapſe of the lungs was neceſſary in the fœtus, it was indiſpenſable for its œconomy, that it be furniſhed with a foramen ovale, &c. &c.

&c. to compenſate for the ſmall allowance of blood that is ſent through them.

In drowning, &c. &c. as very little air remains in the lungs after the laſt expiration, the diſeaſe muſt exhibit nearly the ſame phænomena as the fœtus, whoſe muſcles of reſpiration have not been excited to act; for in this caſe, it is nature that effects what we endeavour to attain by art; that is, to remove the collapſe of the lungs, and this by the introduction of a fluid that will give the neceſſary change to the blood.

Haller, Cullen, and others were of opinion that the ſtate of full inſpiration was as unfavourable to the tranſmiſſion of blood through the lungs, as that of expiration; but this ſuppoſition appears to be but ill-ſupported by fact; for there has been the

teſt

teſt of experiment to prove, that when the lungs were *completely diſtended by water*, the blood freely paſſed from the right ſide of the heart to the left, and the action of the heart, under this circumſtance, muſt have been feeble, if compared to that which it exerts in a ſtate of health.

It has alſo been the generally received opinion that where the *motion of the lungs* is by any cauſe impeded, the circulation, from want of *their mechanical action*, is alſo ſuſpended; and it is ſuppoſed by Mr. Kite, that the accumulation of blood which takes place in the right ſide of the heart, from drowning, hanging, and ſuffocation, originates from the ſame cauſe.

"As it is generally agreed," ſays Mr. Kite, "that the ſtoppage of the *motion of the* "*lungs* is the firſt internal efficient cauſe "of

" of death, let us consider the effects " which reason teaches us, must inevit- " ably follow the cessation of that im- " portant action. The blood returning " from all parts of the body by the su- " perior and inferior cava, is collected in " the right auricle and ventricle of the " heart, from whence in a state of health, " it is transmitted through the pulmonary " artery and veins, into the left auricle; " but in the present instance, the *motion of* " *the lungs* being stopt, only a small quan- " tity can pass through that viscus.

This opinion of Mr. Kite's has been contradicted by experiment, which proves that from the mere removal of the collapse, independant of any *mechanical action* of the lungs, the circulation through them was restored; whence it is obvious that the accumulation of blood in the right side of the heart does

does not proceed from want of *motion*, but from the *collapſe* of the lungs.

To the oppoſite opinion, however, Mr. Kite ſtedfaſtly adheres; and in order to ground his aſſertion, that the circulation ceaſes in *drowning, hanging, and ſuffocation*, from want of *motion* in the lungs, and not from their *collapſe*, he has recourſe to analogy, and obſerves, " that in the action of laugh-" ing the lungs are *dilated*, and remain " almoſt in the ſame ſtate until the cauſe " ceaſes; but while it continues, the blood " cannot be tranſmitted freely through the " lungs; hence we eaſily account for the " redneſs and ſwelling of the neck, face, " and head; and if the paſſage through " the lungs is long impeded, the brain " ſuffers, and apoplexy enſues, which has " on many occaſions ended fatally.

"Cafes have often happened of violent "ftraining and fits of coughing, which "are attended with *a full and long conti-* "*nued infpiration*, terminating in the fame "manner; and finging or crying produce "fimilar effects, although it feldom hap- "pens they are carried to any dangerous "excefs. Inftances almoft out of number "might be brought forward in fupport "of thefe arguments; but enough has "furely been faid to fatisfy the doubts "of the moft incredulous, and fix the "wavering mind of the moft incorrigible "fceptic."

If Mr. Kite's affertion were true, that in the act of laughing the lungs are *dilated*, and that coughing, fits of ftraining, finging, &c. are attended with *a full and long continued infpiration*, this I acknowledge would be fufficient to imprefs conviction on on the mind of every sceptic that the want of *motion* in the lungs, from whatever cause, may prevent the transmission of blood from the right side of the heart to the left; but instead of these efforts being, as Mr. Kite states them, acts of *inspiration*, they are all acts of *expiration*, and we might with as much propriety assert, that charging a gun produces the explosion, as that the acts of laughing, coughing, singing, &c. are the effects of *inspiration*. From long continued expirations, as laughing, coughing, &c. when carried to excess, a *collapse* of the lungs must arise, and this, by obstructing the free passage of the blood through them, will occasion an accumulation of it in the right side of the heart, from which apoplexies may sometimes follow. But supposing the lungs were, as Mr. Kite conceives them to be, in a state of *dilatation*, then apoplexy could never be the conſequence, ſince a free paſſage would be then open to the blood, and prevent the poſſibility of its congeſtion in the head; ſo that Mr. Kite's *efficient cauſe* of death here contradicts *his proximate.*

The argument alſo adduced from analogy, to ſupport the opinion, that the *want of motion* in the lungs ſtops the circulation in *drowning, hanging and ſuffocation*, in reality confutes it, and proves collapſe to exiſt. But had not Mr. Kite ſeemed to conceive that his arguments were ſufficient to ſatisfy the moſt incorrigible ſceptic, we ſhould not have taken ſo much pains in endeavouring to diſprove them. Indeed, as the plan of treatment recommended by Mr. Kite muſt be ſo highly detrimental, if *collapſe* does really exiſt; it appeared of the utmoſt conſequence to determine whether from *it* arofe the fufpenfion of the cir-lation, or from the want of *motion* in the lungs.

In drowning and in fuffocation from foul air, anatomical refearch has difcovered that the veins of the head are not more diftended than in natural death; and that apoplexy does not take place as Mr. Kite fuppofes from hanging, is equally obvious; for if fuch were the cafe, never could we be able to effect a recovery, fince our endeavours to remove common apoplexy, even while the procefs of refpiration and circulation proceed, frequently prove unfuccefsful.

Were it really true that apoplexy took place either in drowning, hanging, or fuffocation, we fhould conceive more fanguine hopes of recovery after breathing had ceaſed in ordinary apoplexy than when it aroſe from drowning, &c. for theſe latter cauſes produce their fatal effect in a few minutes; while common apoplexy, even where a prediſpoſition exiſted, is generally many hours, and ſometimes days before death takes place. If, therefore the two diſeaſes be of the ſame ſpecies, that which ariſes from drowning, &c. muſt be much the more violent in degree. Were this indeed literally the fact, we ſhould then from drowning, &c. find great extravaſation, and no recovery could be effected, and we ſhould have reaſon to expect a recovery in every inſtance, where the cauſe was ſo ſlight as to require ſeveral hours to ſtop the natural actions; but as we are able to recover long after breathing has ceaſed in that diſeaſe, which according to this theory, muſt be the moſt violent, and as we frequently fail of recovering from

from common apoplexy, even during respiration, it certainly proves that this diſeaſe, and that which takes place from drowning, are as eſſentially different as any two diſeaſes to which the human body is obnoxious.

It has been advanced by ſome authors, that the mere diſtention of the veſſels, without any extravaſation either of blood or ſerum, is ſufficient to produce apoplexy, and this is the ſpecies of apoplexy which Mr. Kite conceives to be produced in drowning, &c. as it is acknowledged that no extravaſation takes place in the head; but were congeſtion alone, in theſe caſes, the cauſe of death, then muſt it be ſuppoſed that the diſtention alone of the veſſels acts much more violently than when attended with actual extravaſation; but this is an opinion not only diſcountenanced by probability, but alſo flatly contradicted by Valſalva and Morgagni on the ſtubborn faith of numerous facts. The latter obſerves " that thoſe caſes are the " moſt violent, and much the ſooneſt " mortal, which have their origin from " *extravaſation* within the cranium, we " not only have daily proofs of ourſelves, " but it has alſo been frequently obſerved " by others."

It would therefore appear that though the veſſels of the head were fully diſtended in drowning, hanging, and ſuffocation, this diſtention could not here be conſidered as the immediate cauſe of death, ſince at moſt it can produce but a very mild ſpecies of apoplexy; for even when extravaſation follows, the actions of life generally continue for hours, while in drowning, &c. it is needleſs to repeat, the

natural

natural functions are in a few minutes abolished.

There still remains one observation, which proves the impossibility of apoplexy happening from drowning, &c. and that is, that no accumulation of blood can be formed even at the right side of the heart, prior to the commencement of the collapse of the lungs, but as soon as this obstructs the free passage of the blood, then it receives but an imperfect change; and is therefore, in a great measure, deprived of its essential quality. From this circumstance it will no longer be capable of keeping up the full and natural action of the heart and arteries; and as the carotid and vertebral arteries will also have their action proportionably diminished, the impetus of the blood to the head must thereby be checked, and considerably enfeebled. These considerations make it obvious that apoplexy can only happen where the blood receives its proper ſtimulus from the air to ſupport the action of the heart and arterial ſyſtem, and where an obſtruction exiſts to its free return.

In apoplexy that proceeds from diſtention of the ſtomach, and other cauſes, the blood continues to receive its due ſtimulus from the air; while for want of a ſufficient expanſion of the lungs, (the diaphragm not being allowed a proper deſcent,) an obſtruction ariſes to the free return of the blood, which occaſions the diſeaſe. But even in this ſuppoſition, death might not be the conſequence, at leaſt for many hours, if at all; although the veſſels of the head might have been fully diſtended, and that by the natural action of the carotid and vertebral arteries; but as in

drowning,

drowning, &c. theſe veſſels are ſoon deprived of their wonted ſtimulus, no injury whatever can happen to the brain.

From theſe obſervations, we truſt it has been proved not unſatisfactorily, that apoplexy never happens in drowning, &c. but there is an experiment which muſt always ſuperſede argument that fully diſproves the exiſtence of apoplexy.

This experiment has been mentioned before to prove a different fact; but as it is one that ſerves our preſent purpoſe, the repetition of it will therefore be excuſed.

EXPERIMENT.

The trachea of a dog was laid bare, and ſecured by a ligature, and this was endeavoured to be performed at the in-

ſtant

ſtant an inſpiration was made; in leſs than four minutes he ceaſed to ſtruggle. On examining the heart we found the quantity of blood in the left, when compared to that of the right as thirteen are to twelve. A portion of the cranium was removed, and the veins of the head were evidently leſs diſtended than *natural*.

Here then there being no obſtruction to the paſſage of the blood through the lungs, it could not be collected in the right ſide of the heart, and conſequently no accumulation was found in the head, and yet this animal died as ſoon as other animals from ordinary hanging; which carries conviction to my mind, that apoplexy forms no part of the diſeaſe.

As a further teſtimony, however, in favour of this opinion, the following experiment was made.

EXPERIMENT.

The two carotids of a dog were ſecured *, and in half an hour after this operation he was hanged. In leſs than four minutes he ceaſed to move; on removing a large portion of the cranium the veſſels were found much leſs diſtended than in *ordinary death*.

From this experiment it muſt appear obvious, that as the principal ſource of ſup-

* This experiment of tying up the carotids has been made both by Mr. Haighton, and Mr. Cooper, in order to aſcertain the effects, and in every inſtance it appeared to produce no injury whatever to the functions of the animal.

ply was cut off; inſtead of the veſſels of the brain being in a ſtate of congeſtion, the quantity of blood they contained muſt have been leſs than natural, and conſequently no ſpecies of apoplexy could follow. Yet this animal died as ſoon as other animals which had undergone no ſuch operation.

Mr. Kite, " from a variety of circum-
" ſtances, is induced to believe that me-
" phitic air occaſions apoplexy and death
" in two ways; firſt, by affecting the
" nerves of the trachea in ſuch a manner
" as to render the muſcles ſubſervient to
" reſpiration paralytic; and ſecondly, by
" its ſedative property, deſtroying the ac-
" tion of the brain, and nervous ſyſtem."

To the muſcles of reſpiration being rendered paralytic, there are two forcible objections

jections; first, that the nerves of the trachea have no communication with the muscles of respiration; and, secondly, that if the muscles of respiration were paralytic, no recovery could ever be obtained. Yet Mr. Kite in the next page observes, that "several have been "known to have revived spontaneously;" which certainly proves, that the muscles of respiration could not have been in a paralytic state.

As to the latter opinion, that apoplexy and death are produced by the sedative property of noxious air, destroying the action of the brain and nervous system, it can by no means be reconciled to the idea we have formed of apoplexy; for I believe it is generally agreed that apoplexy must happen from *pressure* on the brain; and we might with equal propriety affirm, that tobacco,

tobacco, and other vegetable poiſons, when taken into the ſtomach, (which actually do produce a ſedative effect on the brain and nervous ſyſtem,) bring on *apoplexy*, as that this diſeaſe is the conſequence of the *ſedative property* of mephitic air. Indeed it appears ſomewhat ſtrange, that Mr. Kite, who has paid ſo much attention to apoplexy, ſhould have imagined that this diſeaſe could ever be produced by the immediate effect of any *ſedative*.

We alſo diſſent from Mr. Kite in opinion, that a *full inſpiration* is ever made in foul airs; for although animals when immerſed in ſuch a medium, may have been heard to cry, yet this affords no proof that a full inſpiration has been previouſly made. This is not an uncommon circumſtance in drowning animals,

ſound

ſound being the immediate act of an *ex-piration*; and there can be no doubt but all animals have a quantity of air in their lungs when immerſed in a noxious medium. But as ſoon as the animal on inſpiration becomes ſenſible of its deleterious influence, it endeavours to expire; and to this endeavour an attempt ſucceeds to inſpire, when the ſame ſenſation recurs as permits very little air to paſs into the lungs.

Dr. Crawford's experiments evince, that when an animal is placed in a warm medium, the venous blood becomes nearly florid.

With a view to aſcertain if an animal could be drowned, and the blood in the left ſide of the heart ſtill retain a florid appearance, the following experiment was made.

EXPERIMENT.

A kitten was immerſed in a warm medium, a little above its own temperature, and permitted to breathe under a large glaſs-bell for twenty-four minutes; it was then drowned in the ſame medium.

On opening the cheſt, it was found that the blood in both ſides of the heart was ſomewhat florid, and yet this animal died, which, however, according to Dr. Goodwyn, ſhould not have happened. But why this animal did die, can be readily explained; for the collapſe of the lungs was here of courſe the ſame as in common drowning, and from it aroſe the *immediate cauſe* that ſuſpends the circulation; but there was ſtill another power operating

operating upon this animal to deſtroy life; for from the intenſe heat and denſity of the medium in which the animal was placed, it was compelled to have recourſe to the proceſs of generating cold, in order to reſiſt this exceſſive ſtimulus; and the act of repelling heat invariably renders the powers of the animal leſs ſuſceptible of action: moreover, the power of generating cold by *evaporation*, was here denied. Notwithſtanding, therefore, that the blood in the left ſide of the heart might be florid, yet the ſuſceptibility of action being feeble, the quality of this blood was inſufficient to ſupport irritability.

It is worthy of remark, that in this and in every ſimilar experiment, the heart had leſs action than uſual, although the blood had this florid appearance; which clearly demonſtrates, that much heat diminiſhes irritability, and this effect is probably produced by the quick action which excessive heat invariably excites, and the debility consequent on the endeavours to resist heat. Hence it must appear evident, that although the blood might possess latent heat in abundance, and what in health would have been a proper stimulus, yet from the solids not being susceptible of action, life could not be supported. The *ultimate effect* of all violent stimuli must be that of a *sedative*; thus heat (which is one of the most powerful stimuli in nature) when applied to a certain degree, acts as a stimulus; but if this be carried to excess, the final effect will be extreme *debility* and death. This is likewise the effect of the use of spirituous liquors, &c. a certain quantity will produce a stimulating effect, without diminishing the powers of the animal; but increase it beyond

this

this, and *debility* will be the conſe-quence.

It has been ſeveral times remarked, from the reſult of repeated experiments, that where the collapſe of the lungs was removed after breathing had ceaſed, the circulation went on freely through the lungs, and diſtended the left ſide of the heart; but when the collapſe exiſted, the left was not diſtended, which evidently proves that the *collapſe of the lungs is the immediate cauſe of the ceſſation of circulation*; and not as Dr. Goodwyn ſuppoſes, *the preſence of black blood in the left ſide of the heart*; nor, as Mr. Kite imagines, *from want of motion in the lungs.*

We do not, however, eſtabliſh the collapſe of the lungs as the *proximate cauſe* of the *diſeaſe*; for by the term *proximate cauſe* is generally underſtood, that which on being removed, the diſeaſe ceaſes. If this definition of a proximate cauſe be adopted, then mechanical obſtruction in the lungs from collapſe cannot of *itſelf* be conſidered the *proximate cauſe*; as by the removal of the collapſe, the right ſide of the heart is merely enabled to empty itſelf, and, by the vis à tergo, to produce an action in the left. But before the proceſs of circulation can be completed, the animal muſt be provided with blood poſſeſſing an increaſed quantity of latent heat, as not only the left ſide of the heart, but the whole ſyſtem wants blood of this quality; ſince in the foetal circulation, the change is received before it reaches the heart, and both ſides have a like ſtimulus. As the heart, however, in the adult muſt be the origin of circulation, ſo it is neceſſary that the alteration ſhould be made immediately,

diately, before the blood enters one of these cavities; whereas in the foetus, the heart not being the origin of circulation, the change is given to the blood before it arrives at that organ.

There would appear a more striking impropriety in saying, that *the black blood in the left side of the heart and arterial system was the proximate cause of the disease*, as *this blood* cannot be changed until it has run the course of the circulation, and returned to the lungs; but that cannot be effected without a previous removal of the obstruction formed by the *collapse*, and exciting the left to contract on its *black blood*; and even if the necessary change could be given during the existence of *collapse*, the lungs could not allow a sufficient quantity of blood to pass through them, to keep up the natural functions of the animal.

To us, therefore, the *proximate cause* of that disease produced by *drowning*, *hanging*, and *suffocation*, appears to be *mechanical obstruction in the interior pulmonary vessels from collapse of the lungs, with a want of latent heat in the blood*; for remove this collapse, and induce the necessary change on the blood, and you cure the disease.

Having thus far attempted to establish the proximate cause, we are naturally led to enquire into the usual remedies employed in this disease; and to select such as appear to be the best calculated to produce a salutary effect.

SECTION

SECTION VII.

Effects of emetics in suspended respiration.

THE proximate cause that results from the suspension of respiration in drowning, hanging and suffocation, we have supposed to be mechanical obstruction in the lungs, with a decrease of stimulus in the blood. The remedies employed to remove it are as numerous and different as the theories advanced to explain it; but of them all, emetics, with which we begin, are perhaps the most ineffectual; their administration must even be attended with no inconsiderable injury, if had recourse to before the action of the vital functions is restored, and even then should be regulated by a serious and vigilant regard to particular circumstances.

No ſalutary effects can be expected from vomits, but in caſes where the proceſſes of reſpiration and circulation have been re-eſtabliſhed, and where enquiry informs us that the ſtomach has been overburdened either with food or ſpirituous liquors. In theſe caſes there may be no impropriety in emptying the ſtomach to facilitate the deſcent of the diaphragm in inſpiration; but to *commence* by the exhibition of emetics muſt be highly improper, as the action and energies of the heart, from its ſympathy with the ſtomach, muſt thereby be conſiderably debilitated. And even admitting no ſuch debilitating effects took place, every attempt to empty the ſtomach muſt neceſſarily be futile until the nervous energy be reſtored in a very ſenſible degree, when they may be exhibited to more advantage.

To

To aſcertain, however, with ſome degree of preciſion, the effects of a powerful emetic, the following experiment was made.

EXPERIMENT.

A Puppy was drowned, and after all ſtruggling had ceaſed, one drachm of emetic tartar diſſolved in two ounces of water, was injected into its ſtomach. The lungs were then inflated, and other means of recovery employed, until the animal made an effort to inſpire; ſoon after which it appeared perfectly recovered.

In ſeven minutes from its apparent recovery it began to vomit; in twelve to purge, and continued frequently to vomit and purge for one hour and ſeventeen minutes, when it died.

On

On examining the ſtomach, it was found empty, but without the ſmalleſt appearance of inflammation.

As a recovery was effected in this animal where ſo ſtrong a doſe of poiſon had been adminiſtered, and that without producing any inflammation, it was deemed requiſite to introduce the ſame quantity of emetic tartar into the ſtomach of another puppy during the healthy actions of the animal, in order to determine if the effects were ſimilar.

The experiment was made in the following manner.

EXPERIMENT.

Into the ſtomach of a Puppy of the ſame litter as that of the laſt experiment,

was

was introduced one drachm of emetic tartar, while its natural actions remained unimpaired; in two minutes it appeared faint, in less than four vomited; in eleven purged, and in fifty-three minutes died.

The stomach, as in the last experiment, was found empty, but the *whole internal coat was nearly in a state of gangrene.*

The result of these experiments exhibits a truly remarkable circumstance, that an animal should be drowned, afterwards have poison injected into its stomach, and yet be recovered and continue to live longer than another of the same order and age, that had received the same quantity of poison in full health; it tends however to evince and ascertain one fact, that medicines introduced into the stomach do not produce the same effect when respiration and circulation are *suspended*, as when *these functions* are duly carried on: and this circumstance somewhat accounts for a phenomenon which to me appears extraordinary, that a recovery should sometimes be effected, even after emetics, tobacco, &c. have been administered in quantities sufficient utterly to destroy the life of the same subject, if given in full health.

It may however at first be doubted, whether medicines that possess a sedative property, like tobacco, would not produce their greatest effect on an animal whose powers were weakest, and consequently destroy the irritability of an animal already debilitated by drowning, &c. much sooner than an animal, the vigour of whose powers remained undiminished.

To aſcertain this point the following experiment was made.

EXPERIMENT.

A Puppy of about a fortnight old was drowned, and after all motion had ceaſed, a ſtrong infuſion of tobacco (one drachm to two ounces of boiling water, and ſuffered to cool) was thrown into its ſtomach; the uſual means of recovery were then employed: in fifteen minutes it made an effort to inſpire, and ſoon breathed tolerably well, but in leſs than ten minutes after, it died.

EXPERIMENT.

An equal quantity of an equally ſtrong infuſion of tobacco was introduced into the ſtomach of another Puppy of the ſame age;

age; it immediately fell motionleſs on the ground, and in leſs than four minutes expired.

Theſe experiments ſeem to prove that, whether medicines have a powerful ſtimulant or narcotic quality, their effects are *diminiſhed* in proportion as the powers of the animal are *decreaſed*.

That medicines however do produce ſome effect before reſpiration is reſtored, has been confirmed by the following experiments.

EXPERIMENT.

A ſmall Puppy was drowned, and the cheſt being immediately opened, the heart was obſerved to contract ſtrongly. Six drachms

drachms of laudanum was thrown into its ſtomach, and there followed almoſt an inſtantaneous diminution of the action of the heart.

This experiment was repeated, by injecting white vitriol, emetic tartar, infuſion of tobacco, &c. into the ſtomach, at a time when the heart was expoſed to view; and theſe were alſo found to check the force and frequency of its contractions, but particularly tobacco. As it therefore appears that in this diſeaſe ſympathetic effects continue to ariſe from the application of impreſſions to the ſympathiſing organs, it will at once appear obvious, that any medicine introduced into the ſtomach which is likely to leſſen the power of the heart, muſt be attended with conſequences highly detrimental; and that brandy, on the contrary, or any other warm cordial, which is known

known to increaſe the action of the heart, (probably in theſe circumſtances without diminiſhing its power) ſhould only be employed.

To confirm this opinion, we proceeded to the following experiment.

EXPERIMENT.

A Dog was hanged, and the heart being expoſed to view, one ounce of brandy was thrown into its ſtomach, the actions of the heart were ſoon quickened, and each contraction appeared more forcible than before the exhibition of this ſtimulus.

This experiment we frequently repeated, by increaſing the quantity of ſpirit to ſix ounces and upwards, and it was found that

that ſo large a quantity quickened the actions of the heart extremely, but they were feeble and of ſhort duration.

From theſe experiments, however, we can draw only this inference, that a ſmall quantity of ſpirits here increaſed *both the power and action of the heart*, while a large quantity quickened the *action*, and *exhauſted the powers*. But the analogy will not hold good with the human ſubject in this particular inſtance; for as the ſtomach of the brute is not accuſtomed to receive ſo ſtrong a ſtimulus as that of brandy, its effects will be different *in degree*. Indeed, from obſerving that all medicines produce a leſs effect after reſpiration has ceaſed, than during health; it is probable that ſix, or even eight ounces thrown into the human ſtomach would not increafe the action of the heart beyond its powers, and thus a cordial of fome kind becomes one of the neceffary remedies in this difeafe.

SECTION VIII.

Effects of bleeding.

WE do not consider bleeding as a dangerous remedy in every case of suspended respiration from drowning, hanging and suffocation; and were it possible to take blood from the part where we know it superabounds, bleeding would prove one of the most immediate and efficacious means of recovery.

The right side of the heart has been found to be loaded with blood. This universally obtains in this disease; and we mentioned one or two instances in particular, where we had an opportunity of observing that the heart ceased to act from over diſtention: but that when relieved from a portion of its burden, its contractions were immediately renewed.

If therefore from the right ſide of the heart, while thus in a ſtate of violent plethora, a ſmall quantity of blood could be taken; experiment and obſervation tell that its power and actions would be inſtantly reinvigorated.

But as this lies beyond the reach of art; the taking of blood from any other part of the body can rarely ever be productive of any advantage, as there is ſeldom preſent in the ſyſtem a greater quantity of blood than is neceſſary to the due ſupport of the circulation. The diminution of this quantity muſt conſequently be attended with hurtful effects.

From

From bleeding, therefore, as a general remedy, little advantage can be expected; nor can it be employed with succeſs but in ſuch caſes, where, from an acquaintance with the complexion and habits of the patients, we may preſume that previous to the accident or diſeaſe, a general plethora prevailed.

It may then be ſerviceable to diminiſh the exceſs of blood that loads the ſyſtem; for when the right ſide of the heart has got rid of its preſent burden, if an accumulation of blood preſſes in every direction on the orifices of the two cavæ, and thence on the right auricle, it muſt tend not a little to enfeeble or wholly deſtroy its action.

Mr. Kite obſerves, that in the tonic temperament, every circumſtance concurs which can contribute to the action of the heart and arterial system, and he immediately after adds that " such people are " also in a state very nearly allied to a " plethora ; hence the blood circulates with " such force as to occasion hæmorrhages " from many parts of the body." In this opinion we cannot coincide with Mr. Kite, as the tonic and plethoric temperaments appear very different; for whenever plethora is present, *debility* must be the consequence, there being a greater quantity of blood in the system, than can permit the animal to take on the tonic temperament.

On this state of the body some light may perhaps be thrown, by comparing it with a disease to which young women are frequently exposed, viz. a difference of temperament producing a sup-

ſuppreſſion of the menſtrual diſcharge. The obſtruction may ariſe either from the preſence of too much or too little blood in the ſyſtem; and yet both ſtates may be properly called the atonic, though for the relief in theſe caſes oppoſite remedies be employed; for in both there is relaxation, inactivity and want of power in the ſolids. In the firſt inſtance we bleed and vomit, by which means the heart and arteries, from having a leſs burden to propel through the ſyſtem, will act more forcibly. In the latter caſe we do every thing to increaſe the volume of blood, a due quantity of which increaſes the action, and ſtrengthens the energy of the heart and arterial ſyſtem. By this it would appear a certain quantity of fluids is requiſite to the ſupport of the proper action of the ſolids; but any thing above or below the ſtandard, will produce

 debility.

debility. It in the firſt caſe it may be called *indirect*, in the other *direct debility*.

Bleeding then ſhould be only employed where the fluids appear too abundant. When the operation is to be performed, I concur with Mr. Kite in adviſing the blood to be taken rather from one of the jugulars; not however that we expect with Mr. Kite that much advantage is gained by taking blood from the head after drowning and ſuffocation; but as there is here a nearer connection with the ſuperior cava, the heart would ſooner be relieved, than where it is drawn from the arm.

When blood letting is deemed neceſſary, it is one of the firſt means of recovery to which we ought to have recourſe.

The

The propriety however of bleeding muſt in every caſe be decided by the medical aſſiſtant; and as we cannot take away blood from the heart itſelf, it will moſt frequently be found that the quantity of blood in the ſyſtem is not ſo great, as to impede the circulation, and conſequently the quantity to be taken away ſhould probably never exceed ſix ounces.

Blood letting may with advantage be employed where, previous to the diſeaſe, the heart might be ſuppoſed to have acted more freely from ſuch an operation; but where its powers of action were already feeble, this remedy muſt neceſſarily be productive of infinite miſchief; for if there is not ſufficient blood in the ſyſtem to furniſh a freſh ſupply to the right ſide of the heart at each diaſtole, inſtead of promoting the good effects of the other remedies,

dies, it may totally fruſtrate or at leaſt retard them.

After hanging however, there will be a much more frequent occaſion for blood letting, than after drowning or ſuffocation; ſince the cord muſt in ſome meaſure prevent the free return of blood by the veins; and although we have endeavoured to prove that apoplexy can never happen, yet in theſe caſes as there is more than the natural quantity of blood in the head, it may be of ſervice to leſſen it; but the quantity of blood in the head will much depend on the weight of the patient; and as bulk, weight, and general plethora frequently are united in the ſame perſon, bleeding becomes here indiſpenſably neceſſary; whereas, if the patient be tall and thin, the diſtance from the heart to the head conſiderable, and the ſyſtem rather to want blood, bleeding,

even

even in cafes of fufpenfion, would perhaps do more mifchief, by debilitating the fyftem, than advantage could be gained, by relieving the local plethora of the head and heart; for if the removal of the local plethora tends to increafe the general debility, this laft difeafe is more dangerous than the one we endeavour to remove.

We fhall next enquire into the effects of electricity, together with thofe of artificial refpiration, both fingly and combined.

SECTION IX.

Effects of Electricity and Artificial Respiration.

FROM electricity, as it has hitherto been recommended and employed, confiderable indeed muft have been the mifchief that enfued. Agreeably to the method that was to direct its application, it was to be adminiftered as a local and general ftimulant, to be tranfmitted through every part of the body, the heart, brain and fpinal marrow, and in all cafes where electricity was the remedy principally relied on, it feemed to fuperfede moft of the other curative operations, but particularly that of expanding the lungs. From attending however to the nature of the difeafe produced

duced by ſuſpended reſpiration in drowning, hanging, and ſuffocation, it will evidently appear, that ſtimulating the heart, without at the ſame time endeavouring to remove the obſtruction of collapſe, muſt be one of the moſt ill-judged and moſt dangerous plans of recovery.

I repeat, there is mechanical obſtruction in the lungs from collapſe. This alone points out the danger of ſtimulating the heart, when there exiſts a cauſe that muſt impede its action; we are deſtroying its irritability, without deriving any advantage, as the circulation can go on to no effect, unleſs the obſtruction in the lungs be firſt removed.

We are, by this plan of treatment, abſolutely taking away life.

Mr.

Mr. Kite conceiving *that the ſtoppage of the motion of the lungs* was the immediate cauſe of ceſſation in the circulation, and that the lungs were not in a ſtate of *collapſe*, was led to recommend ſhocks of electricity to be paſſed through the heart, &c. without the lungs being at the ſame time expanded. In his eſſay he adviſes that artificial reſpiration, as well as electricity, ſhould he frequently *interpoſed,* and that when the body is electrified *all the other operations ſhould ceaſe.*

As it has been proved by experiment, that in this diſeaſe the lungs are in a ſtate of *collapſe,* and that the circulation is ſtopped from this cauſe, and not from the want of *motion* in the lungs; it appears obvious, that Mr. Kite's mode of treatment muſt be highly detrimental. Had this gentleman

entertained

entertained the ſmalleſt ſuſpicion of a *col-lapſe* exiſting, I am perſuaded he never would have recommended the ſtimulus of electricity to be applied to the heart during ſuch a ſtate of the lungs; but have concurred with me in opinion, that ſuch a practice was more likely to deſtroy, than reſtore the actions of life. When electricity has been employed, the lungs have ſometimes been firſt expanded and *collapſed*, and ſhocks then paſſed through the heart, brain, and ſpinal marrow, but in this caſe the lungs being alſo contracted, every electrical ſhock muſt diminiſh the power of the heart. Artificial reſpiration is again employed without electricity, but this ſecond effort promiſes leſs probability of ſucceſs than the firſt; for the heart having before received a ſtimulus, ſo great as that of electricity, it is not likely that the minor one viz. that of the mechanical action of the lungs, ſhould have the ſmalleſt effect. And as the heart may not naturally act more than once or twice in a minute, there are many chances to one that theſe contractions do not happen at the inſtant the obſtruction is removed.

Inflating the lungs, and immediately after preſſing the cheſt, is ſaid to be imitating natural reſpiration, but it appears evident, that this mode of proceeding is very improper, if the heart has not been excited to action during the expanſion of the lungs.

Neither is this proceſs an imitation of nature, for in health, the lungs always contain a quantity of air, and we only expel a little, and receive in proportion. But if we diſcharge all the air as ſoon as received, it is probable, that the heart may act, when the

the lungs are contracted, and which action can produce no salutary effect.

Whatever view the operator may have, who pursues this plan of treatment; whether he supposes a change to be produced *in the blood, within the auricle*, or whether he expects to propel the blood within the lungs into the left side of the heart, he will be equally disappointed. For we have observed, that no change can be produced in the trunks of the pulmonary veins; and we have also found that if any alteration in the quality of the blood be made within the lungs, there is not sufficient quantity remaining for their mechanical action to propel this blood into the left auricle.

The advantage we may expect from inflation, is this; *that the right side of the heart may act at the same time the lungs are*

 distended;

distended; but surely suffering them to collapse, as soon as inflated, is very unlikely to ensure success, when the heart has not been stimulated by electricity during the expansion of the lungs. Moreover, as the air can only become vitiated, by the action of the heart propelling blood into the lungs, there appears no necessity of performing a complete expiration after every inspiration, unless electricity has been at this instant employed.

The plan of treatment necessary to be pursued is obviously this. We should *first expand the lungs, and when the collapse is removed, stimulate the heart by a shock of electricity. The heart from this is made to contract, there is a free passage for the blood, and air in the lungs to produce a change*; therefore if any irritability be left in the heart, some blood must enter the lungs. We now perfectly collapſe them, and of courſe this blood will be conveyed into the trunks of the pulmonary veins and left auricle, and the circulation will go on. The lungs are again immediately diſtended, and kept ſo, until another ſhock be paſſed as before.

Here then it is neceſſary, that all the air ſhould be expelled as ſoon as the heart has been made to act; ſince this air may have loſt the greater part of its purity. But as the irritability of the heart is ſeldom ſuſceptible of action, from the ſmall ſtimulus of inflation; this practice can never be proper where electricity is not employed.

It has been obſerved, that the heart of ſome animals in ſuſpended reſpiration, has for a time the power of overcoming this obſtruction by its own ſtimulus, without removing the collapſe; and probably in man, the heart may poſſeſs a ſufficient degree of irritability to perform the ſame functions on being ſtimulated by electricity, agreeable to Mr. Kite's plan. But without conſidering the powerful ſtimulus, required to effect this, and the debility which muſt neceſſarily enſue; let us enquire, what advantages can be poſſibly gained by propelling blood from the right ſide of the heart to the left, during the collapſed ſtate of the lungs. Allowing this could be effected, there is no air in the lungs to produce any chymical alteration on the *quality* of the blood; and were the left auricle and ventricle, in part emptied and again diſtended with a fluid equally foreign to the wonted ſtimulus, their power muſt every time be diminiſhed, and conſequently the right, at each contraction, require a ſtronger ſtimulus to produce the ſame effect; when as the left finding an increaſing difficulty in propelling

its

its contents; the right would be leſs capable of overcoming the collapſe.

This power therefore could only continue for a ſhort time, and during its exiſtence no better effect could be produced from blood paſſing through the lungs *without receiving a change from the air*, than when propelled from any other artery into a vein by friction.

If no electrical machine can be procured, the manner of carrying on artificial reſpiration ſhould be altered; the lungs are to be expanded; and, inſtead of compreſſing the air out as ſoon as received, they are to be kept in a ſtate of moderate expanſion for about a minute; ſo that if the heart acts during this period, there may be no obſtruction to the paſſage of the blood.

To effect this, repeated inspirations are requisite, allowing at each time the superfluous air to escape before the lungs are made to *collapse*, that there may be in some measure a fresh current of air. By this means the surface of the lungs will at each inspiration be thrust against the heart; and if part of its irritability is lost, so that this shall not act as a stimulus; still when the heart does act, there will be air to give the change, and *no impediment to the passage of the blood*.

It was observed that the lungs in ordinary respiration have no active power in propelling blood through them in health. But it seems in the recovery they may assist by their action; for when the heart possesses only power sufficient to send blood within the lungs, without being able to propel it to the left heart, producing an *artificial collapse*

under

under theſe circumſtances will empty the interior pulmonary veſſels of the blood they have received, and excite the left auricle and ventricle to contraction. That the lungs will here produce this effect, there can be no doubt, ſince we find a greater quantity of blood in them when diſtended than collapſed; and hence by compreſſing the lungs, they muſt act upon all the blood they have received ſince the laſt expiration.

Care however ſhould be taken, that the collapſe is never ſuffered to *continue*; for the heart may act at this period and *then* without effect; ſo that the act of inſpiration in every inſtance ſhould be performed immediately after the laſt complete expiration.

During the whole proceſs of the treatment, from the firſt attempt to effect a recovery, the lungs ſhould never be ſuffered to remain collapſed, that other curative means may be employed. Without this precaution, we render abortive all our endeavours to remove the cauſe of the diſeaſe; for this end not previouſly attained, what rational hope or dependance can be placed in the application of any remedy?

Inſtances of recovery have not been wanting where the lungs were not inflated; but in ſuch it muſt be attributed to an unextinguiſhed energy of the living principle, which continued in ſome degree to enable the muſcles of inſpiration to act ſo as to afford admittance to a portion of air.

Does it not appear probable that the difference of ſucceſs which marks the caſes reported

ported by the Humane Society, in which the ſame method of cure was obſerved; may depend in a great meaſure on the heart's acting, or not acting during the *expanſion* of the lungs? Some patients were irrecoverable after reſpiration had been ſtopped for only one, two, and three minutes; whilſt the recovery of others who had remained more than half an hour under water was effected by a ſimilar mode of treatment.

The variation of the degrees of irritability in the ſame order of animals is found to be conſiderable; but it appears improbable, that one ſhould be deſtroyed from a cauſe which, thirty times multiplied is inſufficient to take away life from another apparently under the ſame circumſtances. Having been preſent at ſeveral caſes of drowning, (in the character of ſpectator,) we had occaſion to obſerve one in particular, in which, though the body had not been long under water, yet all the endeavours to reſtore life proved unſucceſsful. The failure of ſucceſs however in this unfortunate caſe was evidently occaſioned by the means and method purſued to obtain a recovery. The ſmoke of tobacco blown up the rectum, frictions, and inflations of the lungs were firſt employed for about ten minutes, when the two latter were ſuſpended to allow the adminiſtration of electricity. This ſtimulus was applied by paſſing ſmart ſhocks through the heart, brain, and ſpinal marrow; in fact the whole body was electrified. The muſcles through which it was conducted contracted powerfully. The ſhocks were repeated with ſanguine hopes of ſucceſs, but the contractions gradually became more feeble, and in about two hours were totally aboliſhed.

Artificial

Artificial reſpiration, with frictions, was again attempted, but to no effect. It is obvious that in this caſe a conſiderable degree of the vital energy was preſent, but abſolutely deſtroyed by the means employed to re-eſtabliſh it; for as the proximate cauſe of the diſeaſe was not removed, every increaſe given to the action of the heart muſt have produced debility. But had the collapſe of the lungs been taken away when the heart had been ſtimulated, far different indeed might have been the effects; no impediment would then have exiſted to the paſſage of the blood through the lungs, and it would have imbibed from the air its neceſſary portion of heat.

Inflating the lungs and electrifying the heart at the ſame inſtant, may at firſt view be thought a difficult and embarraſſing proceſs; but it will be found that proper inſtru-

inſtruments, conſtructed for the purpoſe, will make this as eaſy, if not more ſo, than what are now employed.

It will be neceſſary, however, firſt to conſider the improvements that have been made, and the diſadvantages that ſtill attend them.

Mr. Kite has formed a very compact caſe of inſtruments for the purpoſe of inflating the lungs, but not without their inconveniences. They are directed to be thus employed.

" A proper perſon ſtationed at the head " of the body to be operated upon, paſſes " the appropriated end of a tube into one " of the noſtrils, and ſuſtaining it there " with the fore finger, compreſſes both " noſtrils ſo firmly between the thumb

" and

" and middle finger of the ſame hand, " that no air can paſs otherwiſe than by " the tube, and the other extremity of " the tube being applied to his mouth, he " blows with force through the pipe into " the noſtrils of the ſubject.

" The medical director ſtanding at the " right ſide of his charge, muſt keep the " mouth perfectly cloſed with his left " hand, while with his right, making a " ſuitable preſſure on the prominent part " of the wind pipe, he prevents the air " from paſſing into the ſtomach, till find- " ing the lungs are properly diſtended, " he is to preſs ſtrongly upon the cheſt, " removing at the ſame time his left hand " from the mouth, ſo as to let the air " paſs out; when by this means the lungs " are compreſſed the proceſs is to be re- " peated, that, as far as can be, the

" manner of natural refpiration may be
" imitated."

We have obferved before, that *collapfing* the lungs as foon as diftended, is not imitating natural refpiration; befides it appears evident that air blown from the mouth of another muft be highly improper, as being robbed in fome meafure of its purity; and if a pair of bellows be ufed, it will employ three perfons, one to inflate, another to fecure the noftrils and mouth, and a third to prefs on the cricoid cartilage, and cheft in expiration; and it feems that unlefs all three perform their refpective offices in perfect concord, the artificial refpiration will be very imperfect.

There are alfo two difadvantages attending every inftrument introduced into

the

the noſtrils; firſt, the epiglottis obſtructs the free paſſage of the air; accordingly, part of the air thus repelled enters the ſtomach, which cannot be prevented by preſſing on the cricoid cartilage; for although preſſure applied here may prevent moſt liquids from paſſing, yet ſo ſubtle a fluid as air blown with force may make its way into the ſtomach; not that air is ſuppoſed to produce miſchief from its quality, but from the mechanical effect it muſt have in preventing the lungs from expanding. We know the detriment which reſpiration in health receives from a diſtended ſtomach, by its preventing the proper deſcent of the diaphragm in the act of inſpiration; for the other muſcles not being able of themſelves ſufficiently to enlarge the cheſt, the right ſide of the heart is prevented from acting with its uſual eaſe; and hence a diſtena diſtention of the ſtomach from air muſt be attended with the ſame effect *.

Mr. Hunter has contrived a double pair of bellows with two valves, ſo that one ſhall perform the office of inſpiration, and the other that of expiration, and theſe are adapted to an inſtrument which is to be introduced into the trachea, after bronchotomy has been performed.

This is certainly a moſt excellent contrivance, but from want of portability, they have rarely been employed.

* From want of proper inſtruments I once ſaw the ſtomach, and the whole inteſtinal canal very much diſtended, and a rupture under which the patient laboured, was alſo conſiderably enlarged; but the major part of the air may at any time be diſperſed, by preſſing on the abdomen.

Dr.

Dr. Monro has invented an inſtrument to be introduced into the trachea, in the form of a common male catheter. This is mentioned by Mr. Kite; but its uſe is only recommended on particular occaſions, and it would ſeem that the inſertion of this inſtrument into the windpipe, could not anſwer the purpoſe ſo well as at firſt might be expected; for when introduced, the inferior oriſice would be thruſt againſt one of the ſides of the trachea, and the curve preſſing on the other, would form an obſtruction to the air.

There alſo ariſes a great difficulty in introducing this inſtrument, more eſpecially to thoſe who have not been in the habit of employing it, as no guide can be given, by which we may know whether it be inſerted into the larynx or pharynx; and as the aperture of the latter is ſo much larger than that of the former, it would rather glide into the œſophagus, than into the trachea, and thus inflate the ſtomach inſtead of the lungs. The ill conſequences ariſing from ſuch a miſtake are ſufficiently obvious; and to guard againſt ſo fatal an error the following inſtrument is recommended.

As it has been deemed requiſite to introduce ſome ſtimulating cordial into the ſtomach, a vegetable bottle (Fig. 7.) is contrived for this purpoſe, which is to be attached to the flexible tube, (Fig. 6. at B.) and introduced down the œſophagus, and on this tube is placed a conical piece of ivory, (cc) that is moveable, to ſerve as a director for the introduction of the pipe into the trachea.

The vegetable bottle being filled, the tube is to be inſerted three or four inches into the œſophagus, and the conical piece of ivory is then to be carried onward by the aſſiſtance of the fore-finger, ſo as to cloſe the ſuperior aperture of the œſophagus.

Having proceeded thus far, the tongue is to be brought as forward as poſſible, and the inferior end of the curved pipe (Fig. the 1.) paſſed to the farther part of the mouth, until it meets with the ivory director. The pipe being then brought a little forward, the ſuperior extremity is to be elevated, by which means the inferior will be depreſſed, and with eaſe enter the trachea: for as the entrance of the œſophagus is ſituated immediately behind the larynx, and as the pipe is prevented from entering here by the ivory director, it muſt paſs into the air-tube; ſo that the vegetable bottle, and its appendages anſwer a double purpoſe, that of injecting fluids into the ſtomach, and as a guide to the introduction of the other instrument.

* The pipe for the trachea is much larger and longer than Mr. Kite's, and made nearly on an opposite scale, viz. the great curve is given to the superior, instead of the inferior part; from which results this advantage, that when it is fixed in the trachea, it will be nearly in a strait line with that tube; and for the more easy introduction of the instrument, the pipe is made conical, and that there may be no impediment to the passage of the air, two lateral openings are made at the inferior extremity (B.)

* It may perhaps be adviseable, that the ivory director be continued in the œsophagus during the whole process of the treatment, as this will effectually prevent any air from regurgitating into the stomach.

The

The application of theſe inſtruments can not be ſuppoſed to embarraſs any profeſſional man; if however, any impediment ſhould prevent the inſertion of the pipe into the air-tube, bronchotomy ſhould be immediately performed; but the place, and manner of performing this operation, agreeable to the method generally recommended, do not appear the moſt eligible.

We are adviſed by authors, to begin it by a longitudinal inciſion, three or four rings below the cricoid cartilage, and when the trachea is met with, to divide it between the rings.

The performance of this operation, according to this plan, can ſcarce be attended with danger, when attempted by a ſkilful anatomiſt; but it may be embarraſſing to a medical aſſiſtant, who is obliged haſtily to perform it when perhaps he may not perfectly recollect the ſituation of the veſſels; and it is to be remembered that haſte is always particularly neceſſary on theſe occaſions. Allowing however, that the operation is ably performed, great inconvenience muſt follow from the ſituation of the wound; for in the recovery of the drowned, hanged, and ſuffocated, the head is, and always ought to be, kept a little elevated, the conſequence of which muſt be, that the aperture in the trachea then becoming the moſt depending part, the flow of blood that follows the operation will principally enter it, and thus prevent artificial reſpiration from being properly carried on. This is not a theory founded on hypotheſis, but on facts; as we have ſeen two caſes wherein this accident actually happened.

An other inconvenience attendant on this mode of operating is, that from the trachea at this part being covered by ſo much integuments, the pipe for inflating the lungs, cannot be properly ſecured; and ſhould a recovery be effected, the patient muſt be under the neceſſity of keeping his chin directed conſtantly downward, in order to approximate the cartilages, a poſition that is not only very diſagreeable, but to be continued almoſt impracticable.

In order therefore to render the operation more ſimple, leſs dangerous, and to prevent blood from entering the air-tube; I conceive it more eligible to divide the thyroid cartilage: and that inſtead of the inciſion firſt being longitudinal, and then tranſverſe, both the integuments and cartilage ſhould be cut through longitudinally at once.

Several are the advantages derived from this mode of operating. Firſt, no danger can then ariſe from the want of anatomical knowledge. Secondly, the covering being here very ſuperficial, little blood will be loſt, and the little that does eſcape, cannot get into the windpipe. Thirdly, the curved pipe can be very well ſecured, in order to carry on inflation and collapſe. Fourthly, if our attempts to recover be ſucceſsful, keeping the head naturally erect, will be the beſt poſition to approximate the divided cartilage ; and laſtly, that the recurrent nerves are in no danger of being divided. The only inconvenience to be dreaded from this manner of operating, is that of committing an injury on the ſacculi laryngis, and thus to incommode the voice ; but theſe are ſecured from danger, by cutting through the middle of the cartilage ; and an union will be as completely effected, as if the trachea itſelf had alone been divided.

The ſurgeon ſtanding at the right ſide of the patient, ſhould perform the operation by putting the integuments on the ſtretch with the thumb and forefinger of the left hand, a longitudinal inciſion is then to be made immediately over the thyroid cartilage, into which may be inſerted the curved pipe that was intended to be introduced into the trachea by the mouth.

Whether this operation has, or has not been performed is of little conſequence to the recovery, if an inſtrument be introduced into the windpipe, that is connected with the other apparatus.

To the curved pipe for the trachea is to be fixed one extremity of the flexible tube, (Fig.

(Fig. 2. A); and the other end (B.) be attached to the inſtrument, (Fig. 3. C.) which may be fixed to the nozzle of any pair of bellows.

Every thing being prepared for inflating the lungs, one aſſiſtant is to have the direction of the bellows, and to ſtand at the head of the patient, whilſt the other prevents any air from eſcaping at the noſtril and mouth; or from the aperture, if any has been made in the trachea.

The bellows are now to be employed, until the cheſt is elevated; and the Medical Aſſiſtant, having the electrical machine prepared, is to place one director between the fourth and fifth rib of the left ſide, and the other between the ſecond and third of the right; ſo that the electrometer may diſcharge the jar, and the ſhock be made to

paſs

paſs from the apex of the left ſide of the heart to the baſis of the right.

When the electrical ſtroke has been once more repeated, the aſſiſtant, who has the care of the mouth and noſtrils, is now to remove his hands, and preſs ſtrongly upon the cheſt; the bellows are again to be immediately employed, and another ſhock being prepared, the heart is to be thus ſtimulated twice or thrice, and the lungs collapſed as before.

If the heart retains any irritability, the effect of this treatment muſt be evident; for the collapſe of the lungs being removed, the contractions of the heart are renewed, a free paſſage is opened for the blood, and air is admitted to give it the change. But as the actions of the heart may probably not be ſufficiently powerful

to

to propel the blood completely through the lungs, it becomes neceſſary to have recourſe to the collapſe, in order to effect this. We therefore, after having inflated the lungs, and electrified the heart, preſs upon the thorax, in order to expel moſt of the air contained in the lungs; for ſuppoſing the lungs have received but one ounce of blood from the contraction of the heart, a certain degree of collapſe will get rid of half of this blood; but if the collapſe is increaſed, the quantity of blood that will be acted upon will alſo be greater. This appears therefore a matter of importance, for the greater the quantity of blood that is ſent from the right ſide of the heart to the left, if at the ſame time it has received the wonted change from the air, the greater undoubtedly is the probability of its exciting the left to action, than when only half the quantity is tranſmitted.

If

If *natural respiration* be imitated without ever attending to the *collapse* of the lungs, there can be little probability of success, even should the heart be electrified during the expansion of the lungs. For if the pulmonary vessels are distended with blood by the action of the right side of the heart, without producing a collapse of the lungs, sufficient to enable them to act mechanically in emptying these vessels; there will arise nearly as great obstruction to the action of the heart as when the collapse existed; for the pulmonary vessels must then be emptied as well as distended by the action of the right side of the heart alone, which by this disease is soon rendered so enfeebled, as to be wholly inadequate to such an exertion.

By exhausting the lungs, after the heart has been made to act during inspiration, the

collapse

collapſe will in ſome meaſure ſupply the abſence of powerful action in the right ſide of the heart; for all the blood the lungs have received is by that means carried to the left, by which we not only gain the advantage of ſending blood which has received its due heat from the air into the left auricle and ventricle, but moreover the pulmonary veſſels are again put in a fit ſtate to receive more blood from the action of the right, and even a feeble contraction of the heart will be capable of ſending blood into the pulmonary arteries, though a more powerful one would be inſufficient to propel it into the pulmonary veins and left auricle.

Mr. Field, a very ingenious mathematical inſtrument-maker in Cornhill, has invented an inſtrument (fig. 4.) which may be ſixed to the nozzle of a common pair of bellows for the double purpoſe of inflating

inflating and collapſing the lungs. (For a deſcription of which ſee the explanation of the plate.) But in order to produce this effect, it is neceſſary that the valve hole of the bellows be cloſed by the inſtrument (Fig. 5.) by which means all the air employed muſt paſs through the ſmall aperture (d in fig. 4.) Hence the operation of inflating and collapſing the lungs neceſſarily becomes a ſlow and tedious proceſs, and which may be conſidered as an imperfection in this inſtrument, particularly if the bellows to which it may happen to be fixed be not air-tight, in which caſe the external air will find a ready entrance, and its intention as an air-pump will be defeated.

If dephlogiſticated air were at hand, there can be no doubt but that it would be far preferable to any other for inflating the

the lungs; but to procure it in ſufficient quantities at ſo critical a period is nearly impracticable, we muſt therefore make uſe of atmoſpheric air as pure as can be obtained.

If the jar be not charged to give the electrical ſhock, as ſoon as the lungs are expanded, no miſchief or inconvenience enſues; for we need only ſuffer a ſmall quantity of air to eſcape at the mouth after every inſpiration, and immediately throw freſh in by the bellows; and this proceſs is to be continued for about a minute; when, if the ſhock is not yet ready, we let go the mouth, and empty the lungs. The heart from this may have been irritated by the repeated inſpirations, while the lungs have not been ſuffered to obſtruct the free paſſage of the blood, and a freſh ſupply of air has been introduced to give

it

it the neceſſary change; ſo that if the heart has acted during this period, collapſing the lungs will now convey the blood they have received into the trunks of the pulmonary veins and left auricle. This proceſs is therefore to be purſued where any circumſtance prevents the ſhock of electricity being given as ſoon as the lungs are expanded, or where no electrical machine can be procured; but as the irritability of the heart cannot long be excited to action by the mere diſtention of the lungs, we think it of the higheſt importance that electricity ſhould be employed.

It ſhould however be remembered that every ſhock given to the heart during the collapſed ſtate of the lungs, tends to rob it of its vital power, without promoting in the leaſt the recovery; and let it alſo be repeated, that the lungs from the beginning are never to be ſuffered to remain collapſed for a ſingle minute; as the heart may act at that very inſtant, and in this caſe without effect; for as every contraction is an expenſive operation to the heart, if it has got rid of no portion of its burden, the utmoſt care ſhould be taken that the lungs be expanded at every ſyſtole of the heart; and this can rarely happen from the uſual method of inflating the lungs without at the ſame time ſtimulating the heart. When the heart has been once emptied, occaſional ſhocks may be tranſmitted through other parts of the body (care always being taken that the heart partake of their influence, and that the lungs be expanded); for ſtimulating the extremities, may probably produce an action in the arterial ſyſtem; but it ſhould be ever in our eye that the heart is to be

conſidered

considered as the origin of circulation, and whilſt other parts of the body are electrified, care ſhould be taken that the heart at the ſame time, partakes of the ſtimulus.

In order to compare the difference of the effects produced by electricity on the heart, when the lungs are collapſed, and thoſe that reſult from it, when the lungs are in a ſtate of expanſion, the following experiment was made.

Experiment.

A Cat was ſtrangled, and five minutes after the laſt expiration the cheſt was opened; the lungs were then alternately expanded and collapſed for five minutes, the heart acted rather powerfully, but no alteration could be obſerved in the blood of its two ſides; either as to quantity or quality.

The heart was now electrified by ſmall ſhocks, during the exiſtence of collapſe, and this was continued for five minutes, when upon examination, it was obſerved that its action was evidently leſſened; the left ſide rather more diſtended than before, but the blood was black in both auricles and ventricles.

The lungs were now expanded, and the heart at this inſtant electrified; after two ſhocks had been given, they were collapſed; again expanded and electrified; and this proceſs was likewiſe continued five minutes. On examining the heart, both ſides were now found leſs diſtended, their action quickened, and the blood in the pulmonary

monary veins, left auricle and ventricle completely florid.

The refult of this experiment, not only proves the advantages of the ftimulating power of electricity on the heart, beyond that of *fimple inflation*; but alfo evinces the fuperiority of adminiftering it in the diftended, over the collapfed ftate of the lungs.

Whatever will excite the heart to expel its black blood, and fupply the left fide, and the whole arterial fyftem with blood, that has imbibed its natural heat from the air, muft be the means of cure, the moft efficacious that can be employed; and this laft experiment feems to confirm the opinion, that electrifying the heart during the expanfion of the lungs, and then collapſing them, is the method the beſt calculated to produce this deſired effect.

With reſpect to the electrical machine the more compact, and at the ſame time the more powerful it is, the better; for as the quantity, neceſſary to be applied, muſt be determined by the jar and electrometer, the more ſpeedily it can be filled, the better. The ſize of the jar neceſſary for the purpoſe, ſhould be about thirty inches of coated ſurface; and the electrometer placed a little more than one third of an inch from the jar, the diſtance of which may be gradually increaſed. It is better that the glaſs of the jar be thin, as the ſhock will then be pungent, for if the glaſs is thick, the ſtroke will be large and denſe; and it appears probable, that the pungent ſtimulus would excite greater action in the heart than

than one that is denſe, without being ſo liable to deſtroy its powers.

All that appears neceſſary in theſe caſes, for the purpoſe of applying electricity is the cylinder, a conductor, jar, electrometer; wire and directors, in order to convey the ſhock to the particular parts we wiſh; and all theſe may be compriſed in a box of twenty inches in length, and twelve in width; and as every medical man may have occaſion to make uſe of electricity for other purpoſes, the expence will not be thrown away, even ſhould he never meet with this moſt ſatisfactory employ of it, the attempting or perhaps the actual reſtoring of the apparently dead to life.

There appears no neceſſity for making uſe of the inſtrument invented by Dr. Goodwyn, for the purpoſe of extracting water from the lungs, as thoſe who have recovered from drowning, muſt all have taken in water, without its having produced any remarkable inconvenience; and as the extracting it would take up a conſiderable time, we think it better as ſoon as poſſible to proceed to the diſtention of the lungs.

We ſhall now inquire into the effects produced by the application of warmth.

SECTION X.

Effects of Warmth.

IT has been the uniform opinion of thoſe who have turned their attention and their pens to the ſubject of ſuſpended reſpiration, from drowning, hanging, and ſuffocation; that the application of heat is abſolutely neceſſary, and that it ought to be made with the moſt gradual, and nearly inſenſible increaſes. This idea ſeems to have been ſuggeſted, by attending to the good effects of warmth on torpid animals, and the manner that nature preſcribed, was that of its being applied in the moſt gradual manner; for where the body has been frozen, a ſudden application of heat has

has been found deſtructive, whereas a leſs degree has proved beneficial.

It would be preſumptuous to deny that theſe obſervations and precautions ſeemed well grounded. But it muſt however be confeſſed, that the detection of any ſtrict ſimilitude between the two diſeaſes, would be attended with no ſmall difficulty. In the one, the vital principle is attacked merely by a ſedative power, in the other, it is endangered by a collapſe of the lungs, which not only prevents the free paſſage of the blood, but at the ſame time deprives it of that due degree of heat, which it borrows from the air.

Dr. Goodwyn has particularly inſiſted on this gradual application of warmth, but his plan of treatment does not coincide with our opinion. He obſerves, "that to fa-

" vour the recovery moſt effectually, the " application of heat ſhould be conducted " on the ſame plan nature has pointed out " for torpid animals. It ſhould be applied " very gradually and uniformly, and it " may be raiſed to 98, but not further " than 100. When the body is warmed " uniformly, and the heat of the interior " parts about 98, we direct our attention " to the ſtate of the thorax, and if the " patient make no attempt to inſpire, we " proceed to inflate the lungs." Nor does this practice appear to be altogether in uniſon with the Doctor's own theory on the nature of the diſeaſe, for external warmth can produce no chymical change on the blood; and as he aſſerts that the heart cannot act until a change has been produced, what great expectations can we form of its being attended with ſucceſs. Moreover, this gradual application of heat muſt engroſs no

inconſiderable

inconsiderable portion of time, already too precious, before the external heat can be much increased, and the action also of the muscles of respiration could rarely be restored before that of the heart.

We also are obliged to with-hold our assent from Dr. Goodwyn's opinion, where he says, that whilst the circulation of the blood continues, the temperature of the body may be raised many degrees above the natural standard without inconvenience. To this assertion is opposed the result of Dr. Fordyce's experiments, which prove that upwards of two hundred degrees of external heat of Fahrenheit's scale could not raise the animal heat three degrees; and it may be a doubt, whether *internal* animal warmth can ever be raised to 98 or 100 by the application of external heat, in cases where *life is present*, but where

where circulation and reſpiration are ſuſpended. The warmth of the body in health may be decreaſed many degrees without much inconvenience; but never can be raiſed more than three or four above the natural ſtandard, without producing pernicious effects; which, to guard us againſt, nature has prudently provided two powers of reſiſting heat, while ſhe has given us only one of generating it. We however perfectly agree with Dr. Goodwyn, that warmth is eſſential, and that in its application it ſhould neither be ſuddenly nor irregularly increaſed; but we can on no account deem it allowable, to wait for any *increaſe* of heat in the interior parts, before the lungs are inflated, as it ſeems impracticable to increaſe the *internal* heat, before this end is firſt accompliſhed, unleſs irritability be abſolutely deſtroyed.

To regulate the application of this remedy, it does not appear neceſſary to aſcertain the degree of heat on the external ſurface of the body, and of the rectum, ſince we can always judge of the warmth of the atmoſphere within five or ſix degrees, and as water whilſt in a fluid ſtate muſt have its temperature nearly equal, we only have to be cautious that the warmth of the room be not at firſt much greater. But as it may be ſome ſatisfaction to the Surgeon to know the degree of heat remaining in the body, (ſince the greater the degree of heat, the greater muſt be the irritability;) it may not be improper or unſatisfactory to be furniſhed with a thermometer, and Mr. Hunter's ſeems the only one that is any way adapted to the purpoſe; ſince aſcertaining the heat of any part of the body, except in canals, cannot be of the ſmalleſt utility.

A Ther-

A Thermometer ſeems alſo neceſſary for regulating the increaſe of heat, ſince our ſenſations are more likely to deceive us afterwards than at firſt; and it is of importance that the warmth be not very conſiderable: perhaps 70 degrees of Fahrenheit's ſcale, is as much as ſhould ever be applied, ſince to ſupport any degree above this produces a waſte of ſtrength, which on the contrary we ſhould endeavour to obviate.

Warmth thus applied is certainly highly expedient, and its effects on the ſyſtem are probably theſe, that the blood in drowning, &c. deprived of the greater part of the latent heat it imbibes from the air, becomes inſufficient to ſtimulate the ſolids; but by the application of ſenſible heat to the ſurface of the body, the heat of the animal is prevented from

being

being ſo ſoon carried off; and thus in ſome meaſure ſupplies the place of that latent heat which naturally is abſorbed by the blood; for although heat be abſorbed from the air in a latent form, it is given out to the ſyſtem in a ſenſible one. Let it not, however, be underſtood that warmth is to effect a cure of itſelf; for we have repeatedly mentioned that the collapſe of the lungs has cauſed an obſtruction to the paſſage of the blood; and before circulation can go on, this obſtruction muſt be removed, and the blood furniſhed with its uſual ſtimulus and change.

Various are the modes of applying heat to the body; warm bath, warm grains, &c. but theſe are means more eaſily directed than procured or put in execution; and there is only one advantage attending them, that of applying heat

more

more univerſally. Even this is counter-balanced by a greater objection, as it prevents us from having recourſe to frictions, and permits ſuch a length of time to elapſe before either warm bath, or grains can be procured.

The more adviſeable method therefore may be, to place the patient on a mattreſs or bed at a proper diſtance from the fire, where every other operation, that is thought proper, can be carried on at the ſame time; and the readineſs with which warmth can be thus applied, muſt certainly be a convenience.

We propoſe next to enquire under what circumſtances frictions may be uſeful.

SECTION XI.

Effects of Frictions.

IT is with great propriety Mr. Kite has limited the uſe of frictions; at the *commencement* of the curative operations they muſt be productive of infinite miſchief: for the right ſide of the heart being already overloaded with blood, we are by the uſe of frictions increaſing its quantity: and it ſcarce can be doubted but that this practice has contributed in many inſtances to fruſtrate the moſt ſucceſsful treatment, by producing an over diſtention, and conſequently indirect debility of the right ſide of the heart. With a view to aſcertain by experiment the effects of early frictions, the following one was made.

EXPERIMENT.

A Cat was ſtrangled; and after it had ceaſed to breathe, the body and extremities were thoroughly rubbed for ten minutes, the cheſt was then opened. On examining the heart, the right ſide was found *fully* diſtended, and the left rather more ſo than uſual, without any ſign of action in either.

An opening was then made in the inferior cava, ſo as to let out a portion of blood; and the action of the right ſide of the heart was ſoon renewed.

This experiment was repeated, and it invariably reſulted, that the more the right ſide of the heart was diſtended, the weaker was its action, and that by letting out a quantity of its blood, the action was reſtored; and where no action was evident, during the diſtention, it was generally renewed by removing part of the blood from the heart.

It is however with friction as with electricity; if made uſe of at one time, it may tend to deſtroy life, and at another it may greatly aſſiſt in the recovery.

In our ſurvey of the common effects of ſuſpended reſpiration, it was obſerved that the aorta and arterial ſyſtem contained a quantity of blood; this point being aſcertained, and it being likewiſe known, that the action of the aorta and arterial ſyſtem is ſuſpended from a decreaſe of the due ſtimulus in the blood, and that the veins have

have little or no contractile power of their own; when once the right ſide of the heart has been enabled to rid itſelf of a portion of its contents by the plan mentioned in the eighth ſection, we ſhould then proceed to frictions, as a ſubſtitute to the natural action of the arteries in health, viz. that of propelling the blood onward, and producing a *vis a tergo* on the blood in the veins. The right ſide of the heart being thus in part emptied, is again pretty rapidly diſtended by the application of frictions, which ſhould be continued as long as electricity is employed; but when from any cauſe we are prevented from electrifying, we ſhould be ſparing and cautious in the uſe of frictions, leſt by over diſtention we deſtroy the action of the heart. From frictions made uſe of as a ſtimulant, little or no advantage can be expected.

The excoriations produced by the application of ſalt, brandy, volatile alkali, &c. muſt be exceedingly troubleſome after recovery; this objection however ſhould have but little weight were any real advantage derived from their uſe; but the application of ſtimuli to the eyes, noiſes to the ear, acrid liquors to the tongue and palate, ſternutatories to the noſtrils, ſcarifications to the ſkin, and the actual cautery, are not only horrid in the very idea, but muſt undoubtedly contribute to extinguiſh the little that remains of animal life, rather than to rouze or re-eſtabliſh it into action; for their effect on the nervous ſyſtem muſt be ſimilar to that of electricity when applied to the heart during the collapſe of the lungs, viz. the deſtruction of irritability. The idea that ſuggeſted ſuch applications muſt have ariſen from ſuppoſing the animal powers to be only in a ſtate of torpor, without

out conſidering that there exiſted a cauſe, without the removal of which all theſe attempts muſt not only prove fruitleſs and abortive, but even deſtructive to life much ſooner, than if no remedy at all had been employed.

There appears to be an objection to the uſe of vitriolic acid with oil, or any application that produces an unknown and partial degree of heat. It may be preferable as a medium for friction, to make uſe of a little common oil or lard, which will rather prevent than occaſion excoriations, at the ſame time that it anſwers every other purpoſe and intention; for the principal end to be obtained by frictions is by means of their mechanical action, and any medium that will facilitate this, appears preferable to thoſe applications which ſtimulate and generate heat, for as much warmth as is deemed requiſite may be applied to the body by more certain and leſs diſagreeable means. Nor ſhould it be forgotten that the circulation even in health, is moſt languid at the remote parts of the body; conſequently the frictions ſhould be chiefly applied to the upper and lower extremities, and the body ſhould be occaſionally rubbed, where it does not interfere with the electric ſhock.

We ſhall next examine into the effects of Enemas.

SECTION

SECTION XII.

Effects of Enemas.

AS tobacco thrown up the rectum in the form of ſmoke was one of the firſt remedies employed in ſuſpended reſpiration, and as we ſee, to our great regret, that it is ſtill too frequently made uſe of, we ſhall endeavour, by a few animadverſions on its effects, to proſcribe its continuance.

Mr. Kite, I believe was the firſt who reprobated the uſe of tobacco; and the arguments he adduces in ſupport of his opinion are truly ingenious.

The

The hiſtory of medical errors, ſcarce affords an inſtance of a more blind and obſtinate prejudice, than that which ſtill induces us to adopt a mode of practice ſo obviouſly deſtructive. It is actually exhibiting a poiſon, that acts as moſt other vegetable poiſons do, by producing ſuch an extreme degree of debility as no powers of life can ſupport; and there can be ſcarce any raſhneſs in affirming that ſuch quantities of tobacco have been adminiſtered in this diſeaſe in the form of ſmoke, as would have exhauſted the vigour of a healthy horſe. And indeed can there be any thing more evidently improper than ſuch a practice? We might with as much propriety recommend tobacco in ſyncope, or in a typhus fever, as in ſuſpended reſpiration from drowning, &c. nor can there be the leaſt doubt entertained of the effects it would produce in either of these diseases.

When we consider the effect that a drachm of tobacco in infusion has upon the system, when given solely to produce a temporary debility in hernia, &c. one would scarcely credit that any person acquainted with this effect, could even think of administering eight or perhaps twelve times this quantity, when the powers of life are reduced to their lowest ebb. It is really an indelible stigma on the profession, that while we cannot but observe the deleterious tendency, even of a small quantity of it, on one disease where we wish to reduce the strength; we nevertheless employ it by wholesale in another, where scarce a spark of life remains unextinguished; with headstrong inattention we

have

have perſevered in its uſe, without ever aſking ourſelves this neceſſary queſtion— What are we rationally to expect from ſuch a remedy? This, indeed, is quackery in the higheſt degree.

When examining the effects of medicines thrown into the ſtomach after reſpiration had ceaſed, it was found that their action was far leſs powerful than when adminiſtered in full health; and it is a fortunate circumſtance indeed, that their operations are regulated by ſuch a law; for if medicines produced the ſame effect in this diſeaſe as during the unimpaired vigour of the natural functions, it may without heſitation be declared, that no one could ever have been recovered where tobacco had been employed in quantities equal to what has been recommended. Tobacco injected into the ſtomach will of courſe produce more violent effects than when thrown up the rectum; but when the quantity employed is perhaps equal to two ounces, the effects muſt be as violent, if not more ſo, than a ſixteenth part injected into the ſtomach.

In order more accurately to determine the effects of tobacco enemas, the following experiment was made.

EXPERIMENT.

A full grown cat was drowned, and the cheſt being immediately opened, the heart was obſerved to act ſtrongly; ſix drachms of tobacco were thrown up the rectum in the form of ſmoke, but before the herb was half conſumed, there remained ſcarcely any action in the heart; and after the whole had been injected, all action ceased to be visible, (without applying the stimulus of electricity.) Mr. Kite has substituted in the place of tobacco, some aromatic herb; if we are to make use of glisters at all, this were certainly preferable, but what great advantages are to be expected from them, is no easy matter to discover.

If the disease is not removed by the means before laid down, we may with as much confidence expect a recovery from injecting a little warm milk and water into the stomach, as from the injection of enemas of any kind.

It should also be remembered that enemas ought to be small in bulk, in order to render them innoxious; for smoke and fluids of all kinds, when given in large quantities,

quantities, will diſtend the inteſtines; the reſult of which will be, that their mechanical effect in preventing the eaſy deſcent of the diaphragm, will neceſſarily be productive of miſchief. Warm enemas may have the ſalutary effect of ſlightly ſtimulating the inteſtines; and the heart alſo from ſympathy, may poſſibly have its action in ſome ſmall degree increaſed, but if tobacco be employed, the oppoſite effect muſt ariſe, and as ſympathy is ſuppoſed to be greater between the heart and ſtomach, than between the heart and inteſtines, it were better to inject ſome warm aromatic, into that viſcus, than into the rectum; but inflation, electricity and frictions, ought by no means to be neglected to make room for ſo ineffectual a remedy.

Having examined the merits of the remedies uſually employed in ſuſpended

reſpiration,

respiration, and recommended such as are countenanced by enquiry and experience, it may not be deemed unnecessary to subjoin an account of the method of conducting the treatment.

SECTION XIII.

Method of Cure.

THE plan of treatment generally to be purſued has been laid down ſomewhat at large in ſeparate ſections, but it may not be unſatisfactory to the practitioner, in theſe caſes, to ſee the whole contracted into an abridged form, and placed in a nearer and cloſer point of view.

As a few minutes in this diſeaſe make a material difference as to the probability of recovery, we think it of ſufficient importance to remark, that the electrical machine and the apparatus for artificial reſpiration, ſhould be kept always at hand, and in readineſs.

As ſoon as we have ſeen the body, we ſhould requeſt that no more ſpectators would be preſent than are abſolutely neceſſary; which we conceive may be eight or nine in all, including the Medical Aſſiſtants; two to have the direction of the cheſt, one to turn the electrical machine, one to direct the ſhock, four to apply the frictions, and the other to aſſiſt occaſionally. This number will be ſufficient for anſwering every purpoſe, and a greater would rather embarraſs, and only contribute to phlogiſticate and render the air leſs fit for reſpiration.

The body, if wet, ſhould be gently dried with cloths, but in ſuch a cautious manner, as to prevent the mechanical effect of the friction from propelling the blood towards the heart.

Having

Having prepared the bed, or mattrafs, on a table of convenient height, the body is to be placed on it with the head a little elevated. Five or fix ounces of brandy, rum, or fome other warm aromatic fhould be thrown into the ftomach, by means of the vegetable bottle and pipe; and the ivory director paffed to the farther part of the mouth, fo as to clofe the fuperior aperture of the œfophagus.

If the patient feems plethoric, and more particularly if the difeafe has been occafioned by hanging; bleeding fhould be employed, and that as one of the firft remedies; nor fhould the application of a proper degree of warmth be neglected.

The curved pipe being then introduced into the trachea, and fecured by an affiftant, and the flexible tube, &c. being at-

 tached,

tached, the lungs ought as ſoon as poſſible to be inflated; and the electrical machine being prepared, one director is to be applied between the fourth and fifth rib of the left ſide, and the other between the ſecond and third of the right; when the electrometer is to be placed a little more than one third of an inch from the jar, and the ſtroke given. The electrical ſhock is to be repeated once or twice, and the aſſiſtant, who prevented the air from eſcaping by the noſtrils and mouth, then ſhould remove his hands, and preſs the cheſt, and immediately after expand the lungs, for the heart to be again ſtimulated.

If any impediment ſhould prevent the introduction of the pipe down the trachea, bronchotomy ſhould be directly performed, in the manner deſcribed in Section the ninth,

ninth, and the curved pipe inſerted into the trachea at this aperture.

When the lungs have been three or four times expanded and collapſed, frictions are to be had recourſe to; theſe, together with the proceſs of expanding the lungs, and at the ſame time electrifying the heart, and then again collapſing them, are to be continued four hours without intermiſſion, unleſs natural reſpiration be reſtored.

In ſome caſes where the living powers are remarkably languid, it may be adviſeable to continue the uſe of electricity, and gentle frictions, even after reſpiration is renewed, as there have been inſtances of momentary and tranſient recoveries: the ill ſucceſs of which may be conceived to ariſe either from the heart not poſſeſſing ſufficient irritability to carry on the circulation, or from want of a supply of blood to the right side of the heart after it has been once emptied. Both these obstacles may be removed by assisting the heart and arteries to perform their respective functions, after the muscles of respiration have been stimulated to action.

If unfortunately no electrical machine be in readiness, or at hand, the method of performing artificial respiration should be altered. When the lungs are expanded, the assistant, who has the charge of the mouth and nostrils, suffers a small quantity of air to escape, while the other assistant continues to throw in a fresh supply: this process should be protracted for about a minute, when the hand is to be removed from the mouth, and the chest pressed, to complete the collapse. It cannot be too frequently inculcated,

culcated, that the lungs are never to be ſuffered to *remain* collapſed; for all our endeavours and attempts to effect a recovery, ſhould the lungs be permitted to continue in that ſtate, muſt ultimately prove fruitleſs and ineffectual.

We cannot better conclude the preſent diſſertation, than by briefly recapitulating the principles and obſervations which form its baſis and ſupport.

CONCLUSION.

FROM what has been obſerved it appears,

1. That during the act of drowning the animal emits air from its lungs, and in its attempt to inſpire, a ſmall quantity of water enters the lungs and ſtomach.

2. That, after the laſt expiration, the lungs are found nearly collapſed, containing a ſmall quantity of froth, but very little air.

3. That the quantity of blood found in the right ſide, is nearly double that contained in the left.

4. That

4. That the blood contained in both ſides of the heart is of the colour of venous blood.

5. That, whether the heart be examined during its contractions, or after they have ceaſed, no perceptible difference is found in the proportions.

6. That the action of the heart ſurvives the periſtaltic motion of the bowels.

7. That the veſſels of the head exhibit no extravaſation, nor even diſtention.

8. That where reſpiration is ſuſpended, from ordinary hanging, the animal has the power of expelling air from its lungs.

9. That although the muſcles of expiration perform their office, no power can be applied to open the trachea to admit air.

10. That as no air can be received, the animal dies with the ſame collapſe of the lungs from hanging as from drowning.

11. That the quantity of blood in the two ſides of the heart bears nearly the ſame proportion in hanging as in drowning.

12. That there is very little difference in the continuance of the irritability of animals after hanging from its continuance after drowning; but the veſſels of the head are ſomewhat diſtended in the former.

13. That animals immerſed in impure air do not appear to make a full inſpiration,

tion, but like animals immerſed in water reject it, as ſoon as a ſenſation is produced in the trachea, which ſeems to make them conſcious of not being in their uſual element.

14. That the muſcles of expiration continue to act till they have expelled all the air from the lungs, which they have the power of acting on.

15. That the ſame collapſe of the lungs is produced from ſuffocation, as from drowning or hanging; and the contents of the right ſide of the heart bear nearly the ſame proportion to thoſe of the left.

16. That animals deſtroyed in impure air are ſooner deprived of their irritability than when reſpiration is ſuſpended from drowning or hanging.

17. That

17. That animals deſtroyed by nitrous air ſoon grow ſtiff and inflexible, ſometimes even before the heart has ceaſed to vibrate.

18. That the veſſels of the head contain leſs blood after ſuffocation from impure air, than after hanging.

19. That in *ordinary* reſpiration and circulation, the lungs are paſſive.

20. That the principal advantage derived from reſpiration, is that of its being the ſource of animal heat; and this heat, by being evolved in a ſenſible form, keeps up the irritability of the whole animal.

21. That the blood imbibes leſs or more latent heat, in proportion to the degree of ſenſible warmth applied to the ſurface of the body.

22. That although the temperature of the florid blood in the left ſide of the heart be at firſt lower than that of the right; yet its ſenſible heat ſoon becomes greateſt.

23. That this circumſtance favours the idea of heat being abſorbed from the air in the act of reſpiration.

24. That as ſoon as the blood has undergone the change in the lungs, it is rendered fit to ſupport the heat and irritability of the animal.

25. That heat is not only evolved from the blood as it paſſes through the capillaries, but that the ſame proceſs continues throughout the whole circulation.

26. That the ſtimulus which excites the heart to act, is the ſame in all its cavities; and this principally is diſtention.

27. That

27. That in the fœtus both ſides of the heart act from the ſtimulus of black blood.

28. That the intent of the fœtal and adult change is the ſame, viz. that of ſupporting animal heat and irritability.

29. That this change is effected in the fœtus, by the blood paſſing through the cells of the placenta, and the veſſels coming in contract with the maternal arterial blood.

30. That ſo much phlogiſton * is imparted to the maternal from the fœtal blood, and *only* ſo much latent heat

* It ſhould have been remarked before, that, whether the doctrine of phlogiſton be eſtabliſhed or not, the theory of animal heat being derived from reſpiration, may ſtill be ſupported, as every phenomenon reſpecting this doctrine can be equally well explained without employing the term phlogiſton.

evolved

evolved from the maternal to the fœtal, as is neceſſary to reſtore the equilibrium of heat and phlogiſton.

31. That as the fœtus is ſurrounded by the warm medium of the liquor amnii and mother, very little heat can be conſumed, and therefore an abſorption of heat equal to that of the adult is not neceſſary.

32. That the fœtus only poſſeſſes one power of reſiſting heat, and as the heat to be imbibed by the fœtal blood is always limited, and as it is always ſurrounded by an uniform degree of temperature, the fœtus ſtands in no need of the power of reſiſting heat, or generating cold by evaporation.

33. That the fœtal heart contains only a ſmall portion of blood that has been to the placenta;

placenta; and as this blood can receive only a partial change, and as even the greater part of that ſame blood muſt firſt paſs through the capillaries before it arrives at the left auricle and ventricle; morever, as that which does not paſs through capillaries mixes with venous blood, it follows that the left ſide of the fœtal heart contracts from the ſtimulus of black blood.

34. That as all the blood which paſſes through the lungs muſt enter the left auricle, the latent heat of the fœtal blood in the right ſide muſt exceed that of the left.

35. That the blood in the umbilical arteries which is to receive the change, being of the ſame quality with that in the left ſide of the heart, is an additional proof that this blood muſt be black.

36. That

36. That although the blood in the fœtal heart and arteries be black, yet, like the blood of the adult in the right ſide of the heart and pulmonary artery, it muſt ſtill poſſeſs a portion of latent heat, which it continues to evolve, in order to keep up the temperature and irritability of the whole animal.

37. That in ſuſpended reſpiration from drowning, &c. the right ſide of the heart continues to act after the left has ceaſed.

38. That the reaſon of this difference is not that the left ſide of the heart is incapable of being ſtimulated by black blood; but from this blood being eſſentially different in quality from that of the right.

39. That this difference of quality in the blood of the left ſide of the heart

depends on its having paſſed through the lungs, and imparted to them a conſiderable portion of its heat, without receiving a ſupply from reſpiration; while the blood of the right poſſeſſes a quantity of heat in a latent form, which it ſtill continues to evolve.

40. That as the blood in the right ſide of the heart, contains a portion of latent heat, while that of the left is exhauſted; and as the ſenſible heat both of the right auricle and venticle muſt conſequently predominate, its irritability of courſe will likewiſe be greater.

41. That the ſtimulus of diſtention being greater in the one than in the other, will tend to produce a difference of action.

42. That

42. That as the right ſide of the heart poſſeſſes more irritability in this diſeaſe than the left; and as the ſtimulus of diſtention is alſo more powerful at the right ſide than at the left, it will be capable of continuing its action when no effect is produced on the other.

43. That although the heart may derive its heat in health, principally from the blood in the coronary veſſels; yet the blood in the cavities of the heart will be alſo capable of evolving heat, and more eſpecially when ſtagnation takes place in ſuſpended reſpiration.

44. That if the right ſide of the heart poſſeſſed the blood of the left; and the left the blood of the right, the degree of irritability muſt be reverſed.

45. That if the right ſide of the heart in ſuſpended reſpiration, had the irritability of the left, and the left the irritability of the right, we ſhould ſcarcely be able ever to effect a recovery.

46. That as ſoon as the action of the left ſide of the heart is increaſed by the ſtimulus of florid blood, the right alſo acts more powerfully.

47. That this depends on the coronary veſſels being ſupplied with blood, that has received a quantity of heat from the air, and which theſe veſſels diſtribute alike to the right, and the left ſide, and conſequently give an equal increaſe to the irritability of both.

48. That the heat and irritability of the heart, being then the ſame, the ſtimulus

of diſtention will produce an equal action.

49. That the heart can be made to act after reſpiration has ceaſed, from the ſtimulus of electricity, while no action can be excited in the external parts from the ſame cauſe.

50. That as electricity is capable of producing action in the heart, when it has no effect on the exterior parts, and as life actually exiſts at this period. It would lead to moſt pernicious conſequences to conclude that life was totally extinct, from no external action being produced by electricity.

51. That as the difference of irritability in animals of the ſame order, depends

more on the ſpecific ſtate of the ſolids, than on the quantity of heat evolved from the fluids, no deciſive prognoſtic can be drawn of the preſence of irritability, from the heat of the animal being above that of the atmoſphere.

52. That as electricity has been found incapable of producing external action, when the heat of the animal was much above the temperature of the ſurrounding medium, it proves that animal heat and evident irritability are by no means coequal,

53. That although heat and irritability are not coequal, yet the greater the degree of heat, the more will be the irritability of any particular animal.

54. That as the heart is conſidered as the origin of circulation, there is a probability

bility of recovery, fo long as the heart can be made to act; although external irritability may not be manifefted by the teft of electricity.

55. That it will ever be better to have no criterion to judge of the abfence of life, and make ufe of every means of recovery, in every inftance, than to rely on an imperfect and hazardous prognoftic.

56. That when the lungs are inflated foon after the laft expiration, both fides of the heart will immediately act.

57. That this probably proceeds from the irritability of the heart being ftill fo great as to be ftimulated to action by the mechanical irritation of the lungs, as in proportion to their expanfion, will their furface prefs upon the two fides of the heart.

58. That

58. That in ſuſpended reſpiration, from drowning, hanging, and ſuffocation, as the collapſe of the lungs begins, the impediment to the paſſage of the blood through them commences.

59. That when the laſt expiration is made, the interior pulmonary veſſels are collapſed, and contain but a ſmall quantity of blood.

60. That if even a change be produced on the quality of this blood, the quantity is ſo ſmall, that unleſs the right ſide of the heart be firſt excited to action, the motion of the lungs alone will be unable to propel this blood into the left.

61. That by inflating the lungs, we cannot alter the quality of the blood in the

trunks

trunks of the pulmonary veins and left auricle.

62. That the right ſide of the heart can propel blood to the left, immediately after the laſt expiration, independant of the mechanical action of the lungs.

63. That as the heart can perform this function after reſpiration has ceaſed, it appears that the lungs have naturally no active power of propelling the blood onward.

64. That part of the black blood contained in the left auricle and ventricle in this diſeaſe, muſt be propelled through the ſyſtem unaltered, whenever a recovery is effected, and as a quantity of blood of the ſame quality has already paſſed into the arterial ſyſtem, it evidently follows that the

left

left auricle can and does act from the stimulus of black blood.

65. That as an animal when immersed in warm water, may be drowned with its blood somewhat florid, it necessarily furnishes an objection to the opinion, that the action of the left heart ceases from the presence of black blood.

66. That as in drowning, &c. the impetus of blood to the head is checked immediately after the obstruction to its return takes place, no injury whatever can happen to the brain.

67. That if apoplexy did actually take place, we should never be able to bring about recovery after respiration had once ceased, since we frequently fail of removing common apoplexy during the ex-

iſtence of reſpiration ; and in drowning, &c. we find no extravaſation. That as no extravaſation takes place in the head, if apoplexy were to exiſt, it ſhould be ſolely attributed to the diſtention of the veſſels.

68. That as mere diſtention is capable of bringing on only a very mild ſpecies of apoplexy, which does not for many hours, and ſometimes for many days, produce its fatal effect; and as on the contrary, apparent death from drowning, hanging, and ſuffocation, takes place in a few minutes ; it certainly follows that this diſeaſe and apoplexy are as eſſentially different as any two diſeaſes to which the human body is expoſed.

69. That the immediate cauſe of the ſuſpenſion of circulation is not *the preſence of black blood in the left ſide of the heart*, neither

neither is it the *want of motion in the lungs*, but it arises from a *collapse of the pulmonary vessels, which produces a mechanical obstruction to the passage of the blood.*

70. That the proximate cause of the disease may be said to consist in a collapse of the lungs, producing a collapse of the pulmonary vessels, with want of latent heat in the blood, since unless both these be removed the disease will still continue.

71. That the mechanical obstruction from collapse must first be removed, before the chymical effects can take place.

72. That emetics in this disease are improper, before the circulation is re-established.

73. That even then they ſhould only be exhibited where the ſtomach is known to have been overloaded previous to the accident that produces the diſeaſe.

74. That all medicines introduced into the ſtomach, produce a leſs effect after reſpiration has ceaſed, than during the healthy actions of the animal; and that in this diſeaſe, all that appears neceſſary to be done is to inject of ſome warm cordial, ſuch as brandy, &c. into the ſtomach.

75. That as in ſuſpended reſpiration, from hanging, there will ſometimes be a plethora in the head, as well as in the right ſide of the heart, bleeding will then be more frequently neceſſary than after drowning and ſuffocation.

76. That the degree of plethora in the head will greatly depend on the weight and bulk of the ſubject.

77. That as in drowning and ſuffocation the right ſide of the heart only is in a ſtate of plethora, and as this cannot be relieved, this operation ſhould never be performed unleſs it appears that there is too much blood in the ſyſtem for the ſolids to act upon.

78. That when bleeding is judged requiſite, it ſhould be performed on the jugular veins in preference to any other.

79. That when bleeding is deemed expedient, it ſhould be one of the firſt remedies employed.

80. That

80. That ſhocks of electricity paſſed through the heart, brain, and ſpinal marrow, without the collapſe of the lungs being at the ſame time removed, muſt tend rather to deſtroy than reſtore the principle of life.

81. That imitating natural reſpiration without frequently producing a collapſe of the lungs is of little avail; for the diſtention of the pulmonary veſſels occaſioned by the action of the right ſide of the heart, will form nearly as great an obſtruction to the paſſage of the blood, as if the collapſe continued to exiſt.

82. That the uncertainty of ſucceſs which has hitherto attended the caſes reported by the Humane Society, has probably been in a great meaſure occaſioned by the method that was adopted of conducting the artificial reſpiration.

83. That the advantages to be derived from artificial reſpiration, are to procure a contraction of the right ſide of the heart when the lungs are dilated, and by collapſe to excite the left auricle to get rid of a portion of its contents, and ſupply it with blood that has renewed its ſtimulus from the air.

84. That in order thoroughly to accompliſh this end, we are to expand the lungs, and when expanded, to ſtimulate the heart by a ſmall ſhock of electricity; we are then to collapſe them, and again to inflate.

85. That from this mode of proceeding, if any irritability remain in the heart, the right auricle and ventricle will be ſtimulated to act, and propel ſome of its blood into the lungs, where it meets with

a free

a free paſſage, and air to impart to it its due ſtimulus and heat; the blood thus duly changed, will, by means of the collapſe, excite the left to get rid of its burden, and furniſh it with a freſh ſupply endued with the proper ſtimulus.

86. That if no electrical apparatus be in readineſs, we ſhould then alter our method of artificial reſpiration.

87. That the lungs ſhould be diſtended, and after allowing a ſmall quantity of air to eſcape, the inſpiration ſhould be repeated; and this proceſs of ſuffering, after each inflation, a little air to eſcape, (or, in other words, imitating natural reſpiration) ſhould be continued about a minute, when we are to exhauſt the lungs; ſo that there ſhould be but one complete expiration here to ſeveral inſpirations.

88. That the intention of this practice is, that as the heart may poſſibly not contract more than twice or thrice in a minute, the blood may find a free paſſage whenever it happens to act, and a freſh ſupply of air to produce on it the neceſſary change; and likewiſe that theſe ſeveral inſpirations may act as ſtimuli to the heart, while the collapſe helps to remove the blood the lungs have received.

89. That during the whole proceſs of the treatment the lungs ſhould never be ſuffered to remain in a collapſed ſtate for a ſingle minute.

90. That electricity ſhould never be employed on any account without a concomitant expanſion of the lungs.

91. That if the heart be excited to act during the collapſe of the lungs, very little more blood can paſs through them than paſſes in the fœtal circulation, and even this ſmall portion receives no benefit from the air.

92. That the application of warmth is neceſſary, and that when firſt applied ſhould be about ſix degrees above that of the open atmoſphere, if this be below 70 of Fahrenheit.

93. That we are on no account to wait for any increaſe of heat, before we inflate the lungs.

94. That placing the body on a mattraſs or bed at a proper diſtance from the fire, is the beſt mode of applying

warmth; as it neither embarrasses nor prevents any other process that may be found expedient.

95. That the principal advantage to be expected from the application of warmth, is to prevent so much sensible heat being evolved from the blood, and which thus in some measure may supply the defect of the latent heat that should have been absorbed from the air.

96. That frictions made use of as a primary remedy are highly improper, as they tend to destroy the action of the heart, by promoting an over distention.

97. That frictions should never be employed before the lungs have been several times expanded and collapsed.

98. That after the heart has been in part emptied, frictions are abſolutely neceſſary.

99. That a little common oil or lard, as a medium for the frictions, is preferable to either ſalt or ſpirits, or any other ſtimulating ſubſtance.

100. That the principal effect to be expected from frictions, is their mechanical action in propelling blood towards the right ſide of the heart.

101. That tobacco in any form is highly pernicious, and were this medicine to produce ſuch baneful effects in caſes where reſpiration is ſuſpended, as in a ſtate of health; it is more than probable that no one could ever have been recovered where this remedy had been applied.

102. That enemas of any kind are only to be considered serviceable, in as much as they co-operate with more important remedies; and if employed at all, warm stimulating ones should be preferred.

103. That their bulk should be small, lest by their mechanical action they prevent the free descent of the diaphragm.

104. That inflating the lungs, electrifying the heart, collapsing the lungs, and the application of frictions, are to be continued four hours, if our endeavours be not previously crowned with success.

105. That electricity and frictions are to be continued even after respiration is restored, if the powers of life seem unequal to the task.

106. That the final intention of the whole plan of treatment, *is to imitate the natural circulation.*

DESCRIP-

DESCRIPTION OF THE PLATE.

FIG. I.

A large ſilver conical pipe to be introduced into the trachea, either by the mouth, or by an opening made in the thyroid cartilage. A. the inferior extremity; B. two lateral openings for the paſſage of the air; C. the ſuperior end of the pipe.

FIG. II.

A ſhort flexible tube for conveying air into the lungs. A. the inferior extremity to be attached to the ſuperior one of the ſilver pipe; B. the other extremity to be

connected

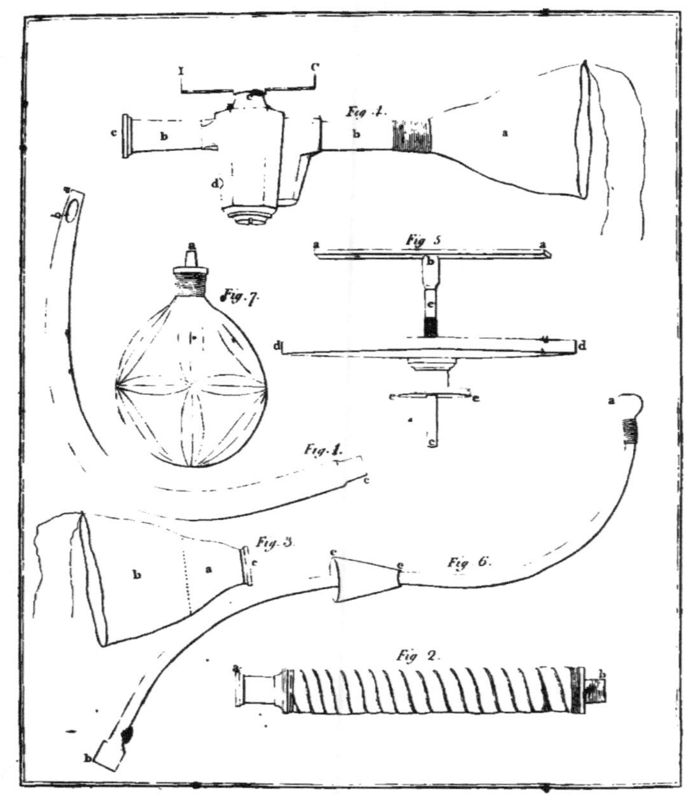
Fig. 1.
Fig 5
Fig. 7.
Fig. 4.
Fig. 3.
Fig 6.
Fig 2.

connected with the contrivance (Fig. 3. at c.) or attached to Mr. Field's instrument (Fig. 4 at e.)

Fig. III.

Represents a short conical brass tube connected with a conical piece of leather, to receive the nozzle of any pair of bellows, and by means of packthread to retain it in its situation. a. the brass; b. the leather portion, c. a female screw to admit the superior extremity of the flexible tube.

Fig. IV.

Mr. Field's instrument for inflating and collapsing the lungs. a. is a conical leather tube to be attached to any pair of bellows; b b. is a brass tube; c. is a stopper to the cock in which there are two valves

valves opening in contrary directions; D. is an aperture through which all the air is to pass to and from the bellows, (the valve of the bellows being previously closed by another instrument represented in the next figure); E. the inferior extremity of the brass tube to be connected with the superior end of the flexible tube, (at B.) When the stopper stands as is here represented with I. (signifying inflation) pointing to the inferior extremity of the tube, the lungs will be expanded, and when the stopper is turned half round, so that C. (meaning collapse) will be placed in the same direction, the lungs will be exhausted. In the one instance by the action of the bellows, air is received at the aperture D. and thrown into the lungs, but prevented from regurgitating on account of the valve.

In the other, air will be received from the lungs into the bellows, and thrown out at the aperture D.

Fig. V.

Is the invention for closing the valve of any bellows. A A. is a piece of iron to be inserted into the valve-hole of any bellows, which being placed across, prevents its being drawn out. B. Is a pivot on which the iron part C C turns. D D. is a circular flat piece of wood, (lined with leather,) to cover the valve-hole, with an aperture in its centre, to admit the iron C C. through it. E E is a brass nut, which is made to screw on the iron C C, to retain the piece of wood in its situation.

Fig. VI.

A flexible tube (of the same composition as flexible catheters) to be introduced

duced into the œſophagus, for conveying ſpirits, &c. into the ſtomach. A. The bulb and inferior extremity. B. the ſuperior. CC. is a conical piece of ivory, to be paſſed a ſmall way down the œſophagus by the fore-finger, to prevent air eſcaping into the ſtomach, and as a guide for the introduction of the ſilver pipe into the glottis, when bronchotomy has not been performed.

FIG. VII.

Is a vegetable bottle, for injecting fluids into the ſtomach through the flexible tube. A. The mouth of the bottle to be attached to the extremity of the flexible tube at B.

FINIS.

www.ingramcontent.com/pod-product-compliance
Lightning Source LLC
Chambersburg PA
CBHW021504210326
41599CB00012B/1134

* 9 7 8 3 3 3 7 8 1 4 3 3 5 *